C.W.

OUR RADIANT WORLD

OUR RADIANT

DAVID W. LILLIE

 IOWA STATE UNIVERSITY PRESS AMES, IOW

WORLD

DAVID W. LILLIE, independent consultant and author on energy and materials, was formerly Liaison Scientist, General Electric Research and Development Center, Schenectady, New York.

Composed by The Iowa State University Press
Printed in the United States of America

First edition, 1986

Library of Congress Cataloging-in-Publication Data

Lillie, David W. (David Waddell), 1917–
    Our radiant world.

    Bibliography: p.
    Includes index.
    1. Radiation—Safety measures. 2. Radiation. I. Title.
RA569.L55   1986   363.1'79   86–15344
ISBN: 0–8138–1296–8

CONTENTS

v

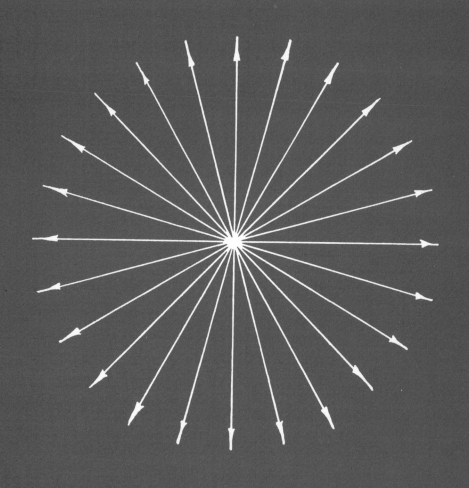

PREFACE

WHEN OOG THE CAVEMAN COWERED at the mouth of his frigid cave watching the lightning flashing outside as the violent storm rumbled by, it never occurred to him that this mysterious frightening flash had within it the embryo of his greatest future boon, fire. Much later some descendant, bolder than the rest, brought back smoldering logs from a forest fire started by that same fearful lightning. Although often burned, he learned to tame the fire and enjoy the heat radiating from it. Light from the sun, moon, and stars, and heat from his fire were the only forms of radiation he could sense and take advantage of. He was totally unaware that other forms of radiation, unfelt and unseen, were part of his everyday environment. Inside his cave the surrounding rock contained tiny amounts of uranium undergoing radioactive decay and generating radon gas giving off energetic particles bombarding his lungs as he breathed. Cosmic radiation from outer space hurled submicroscopic penetrating particles at him, and his own body included radioactive elements continually emitting particulate radiation through his tissue.

Oog and his descendants were never aware of these other radiations, nor did the radiation ever affect their lives, which were usually cut short at an early age by disease or violent death. Only in very recent times has man recognized the existence of the complex radiation environment that surrunds him, and only in recent times has his life span been so prolonged that the very subtle biological effects of radiation became of concern. This concern is still tinged with the same fear and superstition with which Oog in his time viewed fire. In both ancient and modern times much of the fear came from ignorance, and now, as then, greater knowledge permits us to control and defend against the dangers of radiation while utilizing the enormous benefits intelligent use can bring to energy generation, to medicine, and hence to human well-being.

It is important for all of us to gain a realistic perspective on radiation and its relation to our daily life. We accept without question the light that surrounds us and the thermal radiation that keeps us warm. We docilely stand before X-ray machines for diagnostic pictures or therapeutic treatment. Yet we cry with alarm if a nuclear power plant might be sited near us or we use all legal force available, and sometimes illegal force, to prevent a high voltage transmission line from crossing our property. Our perceptions about radiation shape our actions concerning it, and our actions influence the future of important services and functions.

The purpose of this book, therefore, is to make radiation more understandable, to try to separate myth from fact, to recognize and quantify hazard where it exists, but to put risk in proper perspective with the major rewards intelligent utilization of forces generating radiation can provide. In discussing the various types of radiation we will try to answer the following questions: What are they? Where do they come from? How do we measure them? How can we use them? What should we do about them?

The history of man's recognition and use of radiation is a fascinating one. I hope the story of what it is and how it affects us can be made equally so. Let us try.

ACRONYMS, SYMBOLS, AND ABBREVIATIONS

ABCC	Atomic Bomb Casualty Commission
AC	alternating current
AFR	away-from-reactor (in reference to fuel storage)
Ag	silver
ALARA	as low as reasonably achievable
Am	americium
ANSI	American National Standards Institute
Au	gold
B	boron
BEIR	Advisory Committee on the Biological Effects of Ionizing Radiation
BeV	billion electron volts
Bi	bismuth
btu	British thermal unit
BWR	boiling water reactor
C	carbon
c	symbol for the speed of light
CAT	computerized axial tomography
Cd	cadmium
cgs	system of units based on centimeters, grams, and seconds
Ci	curies
Cm	curium

cm	centimeter
Co	cobalt
CPSC	Consumer Product Safety Commission
Cr	chromium
Cs	cesium
D	deuterium
DC	direct current
EFD	energy flux density
ELF	extremely low frequency
EPA	Environmental Protection Agency
EPRI	Electric Power Research Institute
F	fluorine
FDA	Food and Drug Administration
Fe	iron
g	gram
G	gauss
GABA	organic chemical involved in nerve signal transmission
GeV	billion electron volts
GHz	gigaherz
GM	Geiger-Mueller Counter
GNP	gross national product
GSD	genetically significant dose
H	hydrogen
HLW	high-level wastes
HVDC	high voltage direct current
I	iodine
IAEA	International Atomic Energy Agency
ICRP	International Council on Radiation Protection
IDCOR	industry degraded core rule-making group (for evaluation of the source term)
INPO	Institute of Nuclear Power Operations
Ir	iridium
IRPA	International Radiation Protection Association

K	potassium
keV	thousand electron volts
Kr	krypton
kWh	kilowatt hours
l	liter
LET	linear energy transfer
Li	lithium
m	meter
mA	milliamps
MeV	million electron volts
mCi	millicuries
mg	milligram
MHz	megaherz
mil	thousandth of an inch
MIT	Massachusetts Institute of Technology
mm	millimeter
Mo	molybdenum
mR	milliroentgens
mrem	millirem
MRS	monitored retrievable storage
MW	megawatts
mW	milliwatts
N	nitrogen
NCRP	National Council on Radiation Protection
NEMA	National Electrical Manufacturer's Association
NIOSH	National Institute of Occupational Safety and Health
NMR	nuclear magnetic resonance
NPT	Non-Proliferation Treaty
NRC	Nuclear Regulatory Commission
NRC	National Research Council
NSAC	Nuclear Safety Analysis Center
O	oxygen
OSHA	Occupational Safety and Health Administration
Pa	protactinium
Pb	lead

Po	polonium
ppm	parts per million
psi	pounds per square inch
Pu	plutonium
PWR	pressurized water reactor

Ra	radium
Rb	rubidium
RBE	relative biological effectiveness
Rn	radon

S	sulfur
SAR	specific absorption rate
Se	selenium
Sr	strontium
SSPS	satellite solar power station

T	tritium
Tc	technetium
Th	thorium
Tl	thallium
TRU	transuranic wastes
TVA	Tennessee Valley Authority

U	uranium
UHF	ultra-high frequency
UV	ultraviolet radiation

VHF	very high frequency

WL	working level
WLM	working level month

Xe	xenon

Y	yttrium

Zn	zinc

PART 1
RADIATION

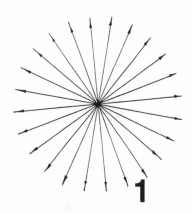

1
WHAT IS RADIATION?

FROM EARLIEST TIMES man has been aware of two kinds of radiation, light and heat. The connection between the two and recognition that they are both only a small part of an enormous spectrum of electromagnetic waves has come in relatively recent times, however. In the late sixteenth century Christiaan Huygens of Holland postulated that light consisted of wave motions in an "aether," the nature of which was rather vague. By the middle of the nineteenth century the mathematical basis for the wave motion of light and heat had been established, largely by James Clerk Maxwell, and gradually thereafter it became recognized that X rays, radio waves, and the gamma rays from radioactive decay were all part of the same continuous electromagnetic wave spectrum. We will see later just how they are related.

But the term radiation as we use it today includes more than electromagnetic wave types. It also encompasses the high energy particles emitted in nuclear or atomic disruptions or changes. The term "radioactivity" is used to categorize this type of event and includes emission of beta particles, alpha particles, neutrons, and also the gamma ray portion of the electromagnetic spectrum referred to above. This distinction between radiation and radioactivity is often a confusing one. "Radiation" is a general term referring to all parts of the electromagnetic spectrum *plus* the particulate emissions from nuclear or atomic rearrangements. "Radioactivity" is the "giving off of radiant energy in the form of particles or rays, as alpha, beta, and gamma rays, by the disintegration of atomic nuclei" (*Webster's New World Dictionary*, Everyday Encyclopedia ed., 1965, s.v.). Thus radioactivity generates radiation, but much radiation exists that does not owe its origin to radioactivity.

3

Radioactivity as a phenomenon giving off radiation was not recognized until 1896. Earlier experiments with cathode rays (later recognized to be beams of electrons) had led to the discovery of X rays by Wilhelm Conrad Roentgen in 1895. These invisible X rays had been shown to produce a visible glow, like that of a luminous watch dial, when they were focused on certain chemical compounds. This was the phenomenon known as fluorescence. It was a fascinating curiosity to many scientists, one of whom was French physicist Henri Becquerel. He began experimenting with certain compounds of uranium that glowed beautifully after a brief exposure to even visible light. Becquerel believed that X rays were being emitted from the fluorescent glow and that he had found a simple way of generating X rays just by shining light on uranium compounds. To prove it he placed the uranium compounds on top of a photographic film, but with the film wrapped in heavy black paper so that the light itself couldn't expose the film. Only the invisible X rays that could pass through the protecting paper could do the exposing. To his delight, it all worked out as he had expected. The film was exposed, and hence X rays must have been emitted out of the fluorescent glow. On 24 February 1896 he reported his findings to the prestigious French Academy of Sciences.

There was only one problem. His conclusion was absolutely wrong. The only saving grace was that he was the one to find that out. Only a few days later he placed a similar uranium compound on a similar paper-wrapped sheet of film in a totally darkened drawer. The film, when developed, showed similar exposure, proving that the rays causing the exposure were being emitted from the uranium compound regardless of exposure to light and their emission was a fundamental property of the uranium itself.

This was the first recognition of the phenomenon of radioactivity. Becquerel's findings began a flurry of work seeking and identifying other radioactive materials. It led to the discovery by Pierre and Marie Curie of the previously unknown radioactive elements polonium and radium and provided the basic clues for unraveling the mysteries of atomic and nuclear structure in the succeeding decades.

Our purpose in this book, however, is not to provide a history of radiation and radioactivity, but to place it in perspective in today's world. These brief historical vignettes only serve to emphasize how short a time exists between our most primitive understanding of radiation and the relatively sophisticated knowledge of today. Even that knowledge still deepens and widens at a relatively rapid pace but largely in areas so complex as to be understandable only to highly trained experts. The fundamentals with which we are dealing are clear and relatively simple, however.

ELECTROMAGNETIC RADIATION

We began this discussion of radiation with some comments on the early recognition of the spectrum of electromagnetic waves. Now it is time to see what the electromagnetic spectrum really is and how we perceive it today. We will limit ourselves to the wave aspects, since these best explain the similarities and differences of the various radiations. Modern quantum theory recognizes a dual particle/wave nature to electromagnetic radiation, but this more advanced and complex interpretation is not essential to us here.

Before discussing the electromagnetic spectrum itself we have to make sure that two fundamental terms used in discussing wave behavior are clear. These are wavelength and frequency. Probably the waves most familiar to us are those seen in water on the ocean or on lakes. If we are on the shore and watch these waves rolling in, we can readily see that they have regular dimensions such as height or distance between crests and that they have forward motion. We can readily count the number of waves passing a given point in a given time, for example. The electromagnetic waves are more regular than water waves but can be described in precisely the same fashion. The number of waves passing a given point in a given unit of time is the *frequency,* commonly described in cycles per second or thousands of cycles per second (kilocycles), since these waves are much smaller than the water waves we are using to sharpen our visualization. The *wavelength* is the distance between corresponding points, say the midpoints, of two successive waves. In the ocean this may be many feet, but in electromagnetic waves it can vary from tiny fractions of an inch to thousands of yards, an incredible range.

All electromagnetic waves travel with the speed of light, roughly 186,000 miles per second (3×10^8 meters per second),[1] enormously faster than our ocean waves. Since the speed in air is essentially constant, it is relatively easy to see that wavelength and frequency vary inversely (i.e., the shorter the wavelength, the higher the frequency). The frequency (ν) multiplied by the wavelength (λ) must always equal the speed of light (c), so $\nu\lambda = c$.

With the concepts of frequency and wavelength in hand we can now look more closely at the entire electromagnetic spectrum as shown in Figure 1.1. This is a simple scale starting with the very shortest wavelength, 10^{-14} meters (which could also be written 0.00000000000001 meters), at the left of the scale and increasing in regular fashion to 10^5 meters or 62 miles at the extreme right. The names we have given to

---

[1] In exponential notation the superscript denotes the number of powers of 10, perhaps more simply thought of as the number of zeros after the 1. Thus 3×10^8 is 3 times 100,000,000 or 300,000,000.

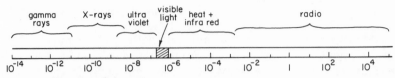

Fig. 1.1. Wavelength in meters.

various ranges of wavelengths are shown above the ranges. The lesson of this chart is that gamma rays from radioactive disintegration, X rays, invisible ultraviolet light, visible light, heat, and radio waves are all part of the same phenomenon and differ only in wavelength and hence in frequency. The names we use for the various parts have been assigned because of special uses or special effects on humans. Almost all parts of the spectrum can be useful to us, and some parts are essential. Similarly almost all parts in too great intensity can be harmful: bright light can cause blindness, thermal radiation can burn, and gamma rays and X rays at high levels can do serious biological damage.

PARTICULATE RADIATION

The other class of radiation that occurs naturally or is generated by man consists of beams of various types of particles emitted from atoms either naturally as in radioactive decay or under provocation such as occurs in nuclear fission. The easiest path to understanding these radiations is to discuss them one at a time. In this chapter we will concentrate on identifying and describing these particulate radiations and in Chapter 2 we will probe more deeply into where they come from and how they are generated.

Shortly after Becquerel's discovery of radioactivity it was recognized that two different types of "rays" were given out by radioactive uranium or thorium. These were distinguished by use of the first letters of the Greek alphabet, alpha (α) and beta (β). Paul Villard in France in 1900 identified a third radiation type from radium that he designated gamma (γ) rays. This nomenclature has remained ever since, unchanged by the enormously greater knowledge we have concerning these radiations today.

We have seen that γ rays were found to be the very shortwave portion of the electromagnetic spectrum, so they will not be discussed further here. To understand the source and nature of α and β particles, we have to see into the structure of an atom and of its nucleus or core. In the simplest view an atom consists of a nucleus surrounded by electrons, tiny charged particles that zoom around the nucleus in mathematically definable orbits. Each electron carries 1 unit of negative electrical charge. Why? Because that's the way we define a unit of electrical charge. It is the amount of charge carried by a single electron, and every electron has exactly the same charge.

Chemical elements differ from one another only in the number of electrons revolving about the nucleus. Hydrogen has 1 electron, helium 2, iron 26, and uranium 92, for example. An atom in its normal state is electrically neutral, however, and since each electron carries 1 unit of negative electrical charge there must be a positive charge on the nucleus sufficient to balance the total charge of the surrounding electrons. The positive charge on a helium nucleus must be 2, therefore, and that for an iron nucleus 26. The positive charge of the nucleus is provided by particles in the nucleus known as protons. These protons are much larger and heavier than an electron. It would take 1,836 electrons to approximately equal the weight of a single proton, for example.

You might think that nature, in order to keep things simple, would have positively charged heavy protons alone in the massive nucleus. That isn't the case, however, because the nucleus contains neutrons as well. The neutrons have almost exactly the same mass or weight as the protons, and, in units known as atomic mass units, both have a value of very nearly 1. A neutron, as the name implies, is neutral in regard to electrical charge. It has none. Therefore a variation in the number of neutrons in a nucleus doesn't change the electrical charge and doesn't change the *chemical* nature. An atom of carbon with 6 neutrons and 6 protons in the nucleus will have the same chemical characteristics as an atom of carbon with 7 neutrons and 6 protons or one with 8 neutrons and 6 protons. The atoms will have different *masses,* however. The nucleus with 6 neutrons and 6 protons will have a mass of 12 (since both neutrons and protons are approximately 1 mass unit), and the atom with a nucleus containing 8 neutrons and 6 protons will have a mass of 14.

The word "isotope" is used to define atoms with different masses and the same chemical properties. We have just described three isotopes of carbon, namely carbon-12 (6 neutrons and 6 protons), carbon-13 (7 neutrons and 6 protons), and carbon-14 (8 neutrons and 6 protons). These can also be written ^{12}C, ^{13}C, and ^{14}C to distinguish one from the other. ^{14}C is radioactive, and we will talk about many other radioactive isotopes in the course of this book.

We return now to the identification of the particles given off by all radioactive isotopes, the particles of radioactivity. An α particle is simply the nucleus of a helium atom. It contains 2 neutrons and 2 protons and has a mass of approximately 4 units. There are no electrons surrounding it, so it has a positive charge of 2 from the 2 protons it contains. Thus it is a heavy, charged particle. Because of its size and charge the α particle has difficulty penetrating into matter, and even highly energetic α particles are stopped by a few centimeters of air. The range of a typical high-energy α particle in biological tissue is only a few hundredths of a millimeter. Our clothes, skin, or other minor barriers protect us adequately from effects of α particles, and thus pure α-emitting radioactive species are only a significant hazard if breathed into the lungs or swallowed.

Beta particles are electrons emitted from the nucleus in the course of radioactive decay. They are the same tiny particles that provide electrical balance to an atom through their unit negative charge of electricity, but they come from inside the nucleus. In view of their light weight they travel at much higher velocities for an equivalent emission energy than do α particles, and largely for this reason have perhaps 100 times as much penetrating power. Their range is still far smaller than that of γ rays that will penetrate perhaps 100 times farther, but it is sufficient to do biological damage. A typical high energy β particle can penetrate about 0.1 centimeter of lead, 1 centimeter of water, and perhaps 10 meters, a bit over 30 feet, of air. Beta radiation can be biologically harmful, particularly if ingested, but can be shielded against fairly readily.

There is one sneaky facet of β radiation that must be mentioned, however. It goes by the term "brehmsstrahlung," a German word meaning "braking radiation." In brehmsstrahlung the final slowing down of the β particle causes the emission of X rays (remember the electromagnetic spectrum?). The phenomenon occurs particularly at high β ray energies and when the β particles are traversing material of high atomic number (those atoms with large numbers of neutrons and protons in their nucleus) such as tungsten or lead. That is one reason why tungsten is used as the target in medical X-ray tubes where the X rays are generated by high intensity electron beams passing into the tungsten and generating brehmsstrahlung X rays. The upshot of this is that β rays may give rise to X rays of much greater range than the β particles themselves, and the resultant X rays contribute in additional fashion to the biologically damaging dose.

The fission of certain elements occurring in nuclear weapons and nuclear reactors is accompanied by emission of very large numbers of neutrons. These neutrons must therefore be included in the catalog of particles existing in radiation environments. Since they have no charge and are relatively massive, they have great penetrating power; it is the existence of these highly penetrating neutrons that contributes most to the need for massive shielding around nuclear reactors. They are not, however, a problem in radioactive waste disposal since, with few exceptions, they are not a product of radioactive decay, only of the fission event itself.

Finally we come to the complicated and fascinating question of particles from cosmic radiation. We'll have more to say in Chapter 2 about the origin and behavior of these particles and for the moment will do little more than catalog what they are. It is now recognized that the earth is under constant bombardment by very high-energy particles originating both in our own solar system and beyond. Most of these particles are protons, the basic building blocks of atomic nuclei mentioned before in the discussion of α particles. In addition, primary cosmic radiation includes high-energy α particles and nuclei of heavier elements such as lithium (Li), boron (B), carbon (C), oxygen (O), and other still heavier particles.

These massive high-energy particles smash into our outer atmosphere where they collide and interact with the oxygen and nitrogen atoms of the atmosphere and produce a great variety of secondary particles including electrons and their positively charged counterparts positrons. The primary particles also produce a variety of secondary particles classed as mesons. These are about halfway in size between electrons and protons or neutrons. The subtle variation of meson types is not important to the main purposes of this volume, and we will not discuss them further. It is enough for now to realize that the particulate bombardment from outer space contributes both in variety and share to the environment of radiation in which we live.

From this brief introduction we can see that each of us lives surrounded by radiation. It takes the form of light and heat, which we sense and understand; of radio waves and X rays, which we cannot sense but for which we have found common uses; and of β particles, α particles, mesons, and heavy nuclei, which we neither sense nor often make use of. Our daily routine—where we live, where we work, how we travel— influences in subtle ways the nature and extent of our interaction with these radiations. Before exploring how this is so, we have to understand some of the terms defining energies and amounts, and we should learn how these are measured.

UNITS OF MEASUREMENT

One of the most important factors characterizing radiation is its energy—how much force is behind it or how great is its penetrating power (although this depends on more than energy alone). The quantity used to describe the energy content of a moving particle or electromagnetic wave is the electron volt (eV). This is a pretty small unit so you will more often see units of thousands of electron volts (keV) or millions of electron volts (MeV) or billions of electron volts (BeV or GeV). The energy released in the fission of a single atom of uranium is about 195 MeV. The very high-energy neutrons emitted from a single fission event will have energy of about 2 MeV, while the so-called thermal neutrons important in most nuclear reactors have energies only up to 0.5 eV (not MeV). Beta particles as emitted in radioactive decay range in energy from 50 keV to about 13 MeV, and γ rays usually fall in the range from 10 keV (.01 MeV) to 10 MeV. You will have to get gradually calibrated as to whether the energy of a given particle or wave is low or high. If an emission is referred to as a "weak β," it is probably in the order of 50 to 500 keV, for example.

The second critical measurement for radiation is the *amount*. How many particles per second are passing through a certain area? How much X ray or γ radiation is striking a given object or person? If you have a dimmer on your light switch over your dining room table at home, it is easy to see that when you turn it up more light falls on the table and when you turn it down there is less light. If you were to be technical about it you could measure in units of footcandles the amount of light reaching the table. The amount of X ray or γ radiation reaching a given area is measured in roentgens. One roentgen is the amount of radiation that produces in 0.001293 grams of dry air 1 electrostatic unit of charge of either polarity (positive or negative). Now that's a pretty scientific and technical sounding definition, but there's a reason for it, and the concept is fairly simple. The only consistent way of measuring and comparing X or γ rays is to measure and compare what they do. We recognize the passage of these radiations because they ionize the air; that is, they knock electrons from the nitrogen or oxygen or other molecules in the air and leave that air electrically charged. If the air were moist, the electric charge would leak away and we wouldn't have a consistent measurement, so we have to specify dry air. It so happens that the weight of 1 cubic centimeter of air when measured at standard temperature and pressure is 0.001293 grams. So what we have really said is that a roentgen is the amount of radiation which, on passing through 1 cubic centimeter of air, will generate one electrostatic unit of charge. The reason it is defined that way is because the electrostatic charge is an easy thing to measure; by measuring the amount of charge produced we are really measuring the amount of radiation received.

When it comes to measuring amounts of *particulate* radiation such as β radiation or α particles there is an even stickier problem. We could merely measure the number of particles passing through a given area in a given time, but that wouldn't let us compare the *effects* of different particles because they might have different energies or might differ in the extent to which they lost energy in the material traversed. Higher rates of energy loss would result in more effect, usually damage, in the absorbing material. We solve that problem by using units describing the amount of radiation in terms of the energy it deposits in the material through which it passes. The first of these units is the rad, defined as the amount of radiation that deposits 100 ergs of energy in a gram of material traversed, an erg being the basic unit of energy in the metric system based on centimeters, grams, and seconds (the cgs system). Note that the rad is a measure of the total amount of radiation received. If you are concerned with the *rate* at which a given amount of radiation is received, you have to speak of rads per minute or rads per hour. The rad is almost always used only for radiation absorbed in biological tissue. It is a fortunate simplification that exposure to 1 roentgen of X rays or γ rays results in about 1 rad of absorbed dose.

A unit with somewhat the same meaning and size but commoner and more useful than the rad is the rem. Its purpose is to permit direct comparison of the biological effects of radiation of different types. A rem is therefore defined as the dose of any ionizing radiation that will produce the same biological effect as 1 roentgen of X rays. Thus 1 rem of β radiation in a certain volume of tissue has the same effect as 1 rem of γ radiation or 1 rem of α particles or 1 rem of cosmic radiation. Some particles have a much more disruptive effect than others over a given range of passage in tissue. For example, 1 rad of α particles with 2 electric charges and 4 mass units is 20 times as damaging when it passes through tissue as is 1 rad (or 1 roentgen) of X rays. Alpha particles are said to have an RBE, or Relative Biological Effectiveness, of 20, which means that 1/20 rad of α particles will give a dose of 1 rem in tissue.

We have now talked about three different units: the roentgen, the rad, and the rem, all used to measure amounts of radiation. The roentgen should properly be used only for X rays or γ rays but frequently is used for other radiation as well, particularly to describe the amount of radiation received in a radiation field or an area where radiation abounds. The rad measures an absorbed dose of any radiation, but when people speak or write about doses of radiation received by an individual over some period of time they usually use rem (or millirem), because the biological effect of a given number of rem is the same regardless of the type of radiation. Remember that all three units refer to a total amount of radiation received over a period of time. The intensity of a radiation dose must be expressed in roentgens or rads or rem per unit of time—rem per hour, for example.

The final measurement term that must be understood is the curie. This is used in describing amounts of radioactive material and is a measure of the number of radioactive disintegrations occuring per unit of time. It was originally the number of disintegrations per second in one gram of radium but is now defined as the amount of radioactive material of any kind producing 3.7×10^{10} disintegrations per second. One curie of a slightly radioactive material might be many grams, while a curie of a very highly radioactive material might be very much less than a gram.

This point leads directly to a discussion of the concept of *half-life* of a radioactive species. It is a characteristic of radioactive material that the amount of radioactivity it emits decreases with time. This decrease is absolutely regular and predictable but is at a different and characteristic rate for each different radioactive species. A given weight of radium-226, for example, will lose half its radioactivity in 1600 years, and in another 1600 years will lose half of the remaining amount, and half of the remainder again in another 1600 years. It is thus said to have a half-life of 1600 years. The reduction in the amount of radioactivity comes about because in 1600 years half of the original radium atoms have disintegrated and through a complex decay chain become stable lead atoms. In another 1600 years half of the remaining atoms will similarly decay. Half-lives of different radioactive species vary from seconds or less to tens of thousands of years. Always keep in mind that long half-lives mean slow rates of disintegration and hence low levels of radioactivity. Short half-lives go with high intensity of radiation. Thus highly intense radioactivity dies away very rapidly, and radioactivity that is with us for very long times is by its nature less intense.

Before leaving the topic of measurement we should reiterate that the quantities described in all of these units vary from very tiny amounts to very large amounts. The electron volt (eV), for example, was mentioned as being a very small quantity, and we frequently talk in terms of thousands of electron volts (keV), millions of electron volts (MeV), or billions of electron volts (BeV or GeV). A curie, on the other hand, is a rather large amount of radioactivity, and we frequently talk in terms of millicuries (mCi) that are thousandths of curies, or microcuries (μCi) that are millionths of curies, or even nanocuries (nCi) that are billionths of curies, and picocuries (pCi) that are trillionths of curies. Thus 1 curie = 1000 mCi = 1,000,000 μCi = 1,000,000,000 nCi = 1,000,000,000,000 pCi. A roentgen or a rem is also a rather large dose of radiation, and milliroentgens (mR) or millirems (mrem) are very commonly used where 1000 mR equals 1 roentgen and 1000 mrems equals 1 rem. Remember to watch for the scale of units being used in articles about radiation. Are they rem or millirem, for example? Table 1.1 summarizes in convenient form what we have been discussing in the previous paragraphs and may be helpful as a reference.

TABLE 1.1.   Units for measurement of radiation

Unit	Abbre-viation	Use	Ranges
Electron volt	eV	Energy of radiation	eV, keV, MeV, BeV
Roentgen	R	Amount of radiation, particularly of X rays and γ rays	mR, R
Rad	rad	Amount of radiation absorbed	mrad, rad
Rem	rem	Radiation dose to humans	mrem, rem
Curie	Ci	A measure of amount of radioactive material present	pCi, nCi, μCi, mCi, Ci

METHODS OF MEASUREMENT

The technology for measuring amounts of radiation is complex and sophisticated, and we will touch on it very lightly here. In considering radiation and how it affects us, however, we should have at least some comprehension of the instruments used, what they measure, and how they work.

First it is essential to understand that there is no single instrument that can measure all amounts of all types of radiation (i.e., β, γ, α particles, neutrons, etc.). Some instruments can be broadened beyond a single function, but none can do all jobs. There is therefore a great variety of different instruments tailored to particular uses.

To simplify the subject we will limit our discussion to field measurements and personnel monitoring. By field measurements we mean measurements of radiation levels existing at some location—perhaps inside a nuclear reactor building, perhaps adjacent to a radioactive sample being shipped, or possibly outdoors where there has been fallout from a nuclear weapons test or some form of nuclear radiation accident. In such an area we want to know the intensity of the radiation, how many roentgens or rems per hour would be received by someone standing at the point of measurement. By personnel monitoring we mean how much total radiation a person has received over the time he or she has been present in an area where radiation may exist.

In an earlier section we mentioned that the passage of X rays or γ rays could be recognized because they ionize the air; that is, they strip off outer electrons and create positively charged atoms. The commonest methods for measuring both X and γ rays simply measure the extent of this ionization and convert that measurement into electrical signals to be read out in the desired units. Such a sensing unit is known as an ionization chamber. In its most elementary form the ionization chamber is a gas (air or argon)-filled metal cylinder perhaps 1½ inches in diameter and 5 to 6 inches long. Down the center runs a small diameter wire insulated at the ends from the chamber wall. A battery is used to apply a voltage between the wire and the chamber. When the presence of radiation causes ionization of the gas within the chamber, an electric current flows between the chamber and the wire. The current in one type of meter, a proportional counter, is proportional to the amount of ionization. In the commoner Geiger-Muller or GM counter every ionizing event regardless of its size causes a large pulse of current that is readily detected and converted to a dial gage reading.

We have noted that γ rays, β particles, and α particles all cause ionization of the air or other gases through which they pass. How can we tell, then, what type of radiation is activating a Geiger counter? The usual solution is to rely on the thickness of the ionization chamber. A stainless steel chamber wall of 15-mil thickness (0.015 inches) will allow all γ radiation to pass through it readily but will exclude all β or α radiation. Tubes as thin as 1.5 mils can be used and will permit penetration by at least some of the β radiation, particularly that which is most energetic. To admit a larger fraction of the β particles or to admit α particles, very thin end windows must be used. Since the range of α particles in air is 1 to 2 centimeters (less than an inch), the counter for α particles must be placed very close to the source if it is to register at all. A thin window counter can discriminate between α, β, and γ radiation by permitting insertion or removal of added thickness at the window. With the thinnest window the counter will register the total ionization from all three radiations. A slightly thicker window will exclude α particles and limit the counter to measuring β and γ only. A still thicker window will eliminate the β particles, and the counter will read only the intensity of γ radiation. Many thin window counters are sold with extra thickness shutters for this simple discrimination.

Typically a GM survey meter will be a metal box about 7 inches long, 3½ inches wide, and 3 inches high weighing about 3 pounds and with a handle to hold it by on top. A scale gage is mounted on top reading in mrem/hr or rem/hr or perhaps mR/hr or R/hr (remember that mrem stands for millirem and mR for milliroentgens), and a lever can be switched between high-sensitivity and low-sensitivity scales. The box contains the battery and electronic equipment that provides the scale calibration. There is a rack, perhaps on top of the handle, for the ionization chamber attached to the box by a flexible cord. In operation the box is held in one hand, usually the left hand, and the ion chamber probe is moved about with the other. The devices are rugged and simple to operate.

Another common configuration is known as a Cutie Pie survey meter. This type of meter has the ion chamber tube mounted in the front of the box, looking somewhat like a telephoto camera lens. The box has a pistol grip underneath, and the gage is mounted on a slanting back panel immediately visible to the operator. This type of unit permits one-hand operation.

Scintillation counters are alternate devices for measuring radiation intensity. They are based on the fact that certain materials, when hit by radiation, give off flashes of light. These light flashes are sensed by an electronic circuit sensitive to light, known as a photomultiplier tube, and the circuit converts the signals into the familiar radiation intensity units. You may wonder how such a device can detect γ rays previously described as a shortwave portion of the electromagnetic spectrum. This comes about because quantum theory has shown us that the wave energy form, while obeying wavelength and frequency laws, distributes its energy via discrete quanta or pulses. These discrete energy pulses are known as photons, and it is these photons that cause the tiny blips of light seen when γ rays strike a scintillating material. Scintillation counters thus can register α, β, or γ radiation with discrimination between them being accomplished by variations in window thickness just as for Geiger-Muller counters.

A variety of light-emitting materials (or phosphors) can be used for scintillation counters, and their special characteristics can help discriminate between radiation types. For example, zinc sulfide (ZnS) with a little added cadmium (Cd) or silver (Ag) responds essentially 100% to α particles but hardly responds at all to β or γ radiation. ZnS is thus used in thin window scintillation counters specialized for α particle measurement. Sodium iodide (NaI) activated by a small amount of thallium (Tl) impurity is used for γ radiation measurement. It is fortunate that the size of the light pulse generated by γ ray photons striking sodium iodide is directly proportional to the energy of the incoming γ ray over a range from 1 keV to 6 MeV. Thus the NaI scintillation counter can tell us the γ ray *energy* as well as *intensity*. A number of organic phosphors such as anthracene, stilbene, or napthalene are also used. These have the advantage of faster response times than the crystals described above. Scintillation counters are in general more expensive than ion chamber types, but have broader capabilities.

The measurement of the intensity of neutron radiation poses some special problems, because neutrons, by their nature, do not cause ionization of the gas through which they pass. The solution is to incorporate some boron in the ionization chamber either as a metallic boron inner coating or as the gas boron trifluoride (BF_3). The neutrons are absorbed in the boron, particularly the isotope boron-10 (^{10}B), which is transformed to lithium with simultaneous emission of an α particle. The α particle ionizes the chamber gas, and the amount of ionization is detected as in regular ionization chambers or GM counters.

In personnel monitoring we want to know how much radiation a person has been exposed to during his work period or stay in an area where radiation is present. This is usually accomplished by film badges that are clipped to one's clothing and contain pieces of film, similar to dental X-ray film, inside a light-tight enclosure. The badge may contain two pieces of film, one more sensitive than the other, to cover a wide range of radiation intensities. There may also be thin sheets of lead or other absorber over part of the film to help identify the type or energy of the radiation. The film can be removed at appropriate times and developed to determine how much exposure has taken place.

There is, of course, a time lag between the time of irradiation and the time the film is finally developed. For this reason pocket dosimeters are often carried as well. These look rather like a large fountain pen and are clipped to or inserted in a pocket. The dosimeter is charged electrically before use so that two fibers repulse one another, and the distance between them can be read on a scale. Any radiation encountered will ionize the gas in the dosimeter and allow charge to leak off and the fibers to come closer together. By holding the dosimeter toward a light and looking through the end, the amount of charge and hence the total radiation received can be read off. This system permits regular checking by the worker or visitor himself if there is any suspicion that he might be in a radiation area higher than desirable.

In recent years a new form of personnel monitoring has come into widespread use involving a phenomenon known as thermoluminescence. Certain crystals after exposure to radiation will emit light when heated to moderately high temperatures (200°C). The amount of light emitted is proportional to the amount of radiation received. A typical personnel badge using this technique will contain a tiny chip of lithium fluoride (LiF) enclosed in a space in the plastic badge. Any radiation encountered by the badge wearer stores energy in the LiF crystal in proportion to the amount of radiation received. At the end of the work period or visit the LiF chip is removed and placed in a scintillation counter with a tiny oven to heat the chip up to about 200°C. The light given off is measured by a photomultiplier tube and is converted electronically into a reading showing mrems of radiation received. This method is simple, convenient, accurate, and cheap.

Sometimes because of an accident or other unusual circumstance it is desirable to have a device that can be attached to one's clothes or easily carried and will give instant and positive warning of excessive radiation levels. There are many types, usually based on ionization measurements, which signal by flashing lights or loud and insistent noise. Some of the most modern solid-state monitors combine audible alarm signals with easily readable numerical displays of the total radiation received or the rate at which it is being received. Alarm monitors of this type can also be permanently located in critical areas to indicate any sudden increase in radiation level that may be hazardous.

There are many sophisticated variants of the types of instruments we have discussed above. The main point to remember is that levels and energies can be measured with great accuracy for all types of radiation. There are instruments sensitive to single particles and instruments that can record the enormous radiation levels seen in weapons tests. No single instrument can do all things, however, and the health physics experts must select the appropriate instrument for the job at hand.

PART 2

ORIGINS OF RADIATION IN TODAY'S WORLD

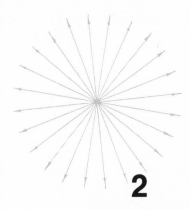

2

NATURAL RADIATION

IN THIS CHAPTER WE WILL EXPLORE the sources of natural radiation in the world—how and where we encounter it, what types there are, and how much of it there is. We'll start with the most familiar, light and heat, and then move on to the less familiar but often as common and all-pervasive forms.

LIGHT AND HEAT

The primary external source of light and heat in our world is the sun, the center of our solar system and its basic energy source. In the sun protons (hydrogen nuclei) fuse to form alpha (α) particles (helium nuclei) with the liberation of enormous amounts of energy. This energy is radiated outward as electromagnetic radiation, and because of filtering effects in the solar atmosphere only the visible, ultraviolet, and infrared portion of the spectrum emerge from the solar surface and travel the great distance to planet Earth.

Here biological species, including man, have emerged with the incredible capability of imaging the visible light through lenses in the eye to provide inputs to the brain that are converted into sensory patterns representing the world around us. You may be thinking that the last sentence is a complex contrived way of saying that human beings can see, and if so, you are correct. Yet it is also true that we accept sight as an obvious response to visible radiation without reflecting on the complexity and the beauty of the process of "seeing." Only in this narrow band of wavelengths are we equipped to sense the radiation so elegantly, even to the point of distinguishing small wavelength differences as color. We can sense infrared radiation as heat, but this is a relatively gross perception, and we sense ultraviolet not at all unless the sunburn response to excessive ultraviolet radiation is counted as sensing.

Light and heat can, of course, be artificially generated on earth. Light can be created chemically via luminescence and flames or electrically via high temperatures as in the filament of an incandescent bulb or via discharges in arc lamps. The fluorescent lamp combines discharge and chemical phenomena to provide light that is uniform and cheap. We can control the intensity of light and its color. We can focus it or diffuse it, and we can use it to magnify images through magnifying glasses and microscopes. Light can be used to improve man's health and well-being. High-intensity lasers permit amazingly precise surgical excisions, and the common sunlamp can give us a cosmetic tan.

Our understanding of thermal radiation helps us use heat precisely to cook our food, warm our bathwater, and control the temperature of our homes. Infrared lamps keep our food warm in better restaurants and bake on the paint that protects our automobiles.

Light and heat have become man's domesticated servants, and only rarely do we see their dangerous side. It does exist, however, for too much light can blind or in lesser amounts cause sunburn and skin cancers. Too much heat can burn and kill. For the most part we understand and respect these radiations and are comfortable with them. Our natural sensors, eyes and skin, enable us to detect the levels of light and heat in our environment and in those rare cases where necessary enable us to take protective measures. It is the unseen, unfelt radiation beyond the visible and thermal spectrum that is fearful to us. This is the radiation we need to come to know better, to get into perspective with the perceived and known.

COSMIC RADIATION

Early measurements with electroscopes had shown that ionization of air existed everywhere and was associated with naturally existing background radiation. Prior to 1910 it was believed that all this sea-level radiation existed because of the radioactive material in the earth. In 1910 a man named Albert Gockel in the United States sent an electroscope up nearly 3 miles in a balloon. He expected the amount of radiation to decrease as the instrument was lifted higher and higher from the earth, but instead the opposite happened and he found evidence of greater amounts of radiation at higher altitudes. This phenomenon was observed again by Victor Franz Hess in 1911 in Austria. He came to the startling conclusion that the observed radiation must be entering the earth's atmosphere from outer space. This was the first recognition of cosmic rays, and for it Hess received a Nobel prize in 1936.

Continued experimentation confirmed Hess's conclusions and also showed that at least some of the cosmic rays had great penetrating power. Cosmic radiation has been detected that has passed through 2000 feet of water, so its initial energy had to be enormous.

The actual origin of cosmic rays is still obscure. Possibly they originate within our own galaxy and are accelerated to their enormous velocities by electromagnetic forces. More probably they originate in space beyond our solar system, perhaps from supernovae explosions or other cosmic events. We now know that the primary cosmic radiation reaching our outer atmosphere is about 89% protons, 9% α particles, and 1% nuclei of heavier elements from carbon up to iron.

This is not what bombards us on the ground, however, since these primary particles, some with energies over a billion electron volts, strike the atoms of nitrogen, oxygen, and other elements in the outer atmosphere and generate a host of new particles plus high energy gamma (γ) rays via these collisions. These secondary particles undergo further collisions creating still more particles and being slowed down in the process. The result of all this is that cosmic radiation is at a maximum about 10 to 15 miles up and gradually decreases as it gets closer to the ground. There it consists of some remaining very high-energy primary particles that have escaped collisions on the way down plus a mixture of secondaries consisting of about 70% mesons, 29% electrons and positrons, and 1% heavier particles. About 40% of the natural radiation received by an individual at sea level is from cosmic radiation, while the remaining 60% comes from radioactive materials in the earth.

It is evident from this that someone in Denver 1 mile up receives a greater amount of cosmic radiation than someone in New York at sea level. Similarly, people flying in a jet airliner at 37,000 feet (7 miles) may receive 60 times as much cosmic radiation in a given time as do people on the ground at sea level. Taking altitude effects into consideration, an average whole-body dose of cosmic radiation in Florida is 35 mrem/yr and in Wyoming is 130 mrem/yr. An airline pilot who averages 20 hr/wk at 37,000 feet for 48 weeks a year would receive about 250 mrem of cosmic radiation above that which he receives in other aspects of his life.

Because primary cosmic rays are charged particles coming towards the earth from outer space, the earth's magnetic field causes additional variations in the amount of primary and secondary radiation that reaches the ground. Cosmic radiation at a given altitude at the equator is only 70% of what it would be at Seattle, Montreal, or London, for example. This variation is not nearly as significant in a practical sense as is the variation with altitude.

NATURAL RADIOACTIVITY

We mentioned in the previous section that about 40% of the background radiation on the ground comes from cosmic radiation and about 60% from radioactivity in the earth. This latter 60% comes from many different origins and varies considerably in different parts of the world. It is the most important part of background radiation to us as individuals because we can to some extent control its level by our own actions and life-style.

Before talking about the specific radioactive atoms that exist naturally in the atmosphere around us it may help to review the concepts of isotopes and the notations used to identify them. Each atomic nucleus is made up of neutrons and protons, the former being electrically neutral and the latter providing the positive electric charge that is balanced by encircling electrons. The number of protons determines the chemical nature of an atom. In other words it identifies the element as being carbon or iron or nickel or uranium, etc. The number of protons thus sets the *atomic number*. For carbon it is 6, iron 26, nickel 28, and uranium 92. Most elements exist, however, in several different isotopic forms that differ only in the number of neutrons in the nucleus. Iron, for example, has four different stable isotopes having 28, 30, 31, and 32 neutrons respectively in addition to the 26 protons that identify it as iron. We said earlier that protons and neutrons have essentially the same mass or weight. The total mass of the nucleus is thus the sum of the masses of the contained neutrons and protons. If we assign a neutron or proton 1 mass unit, then the isotope of iron having 26 protons and 28 neutrons has a mass number of $26 + 28 = 54$. The iron isotope with 30 neutrons has a mass number of $26 + 30 = 56$. The four stable isotopes of iron thus have mass numbers of 54, 56, 57, and 58. We can write the isotope of mass number 56 as iron-56 or more usually, using the chemical symbol, as ^{56}Fe. Some isotopes are inherently unstable, and it is this instability that makes them naturally radioactive. The shift toward more stable forms is accomplished by emission of particles or γ rays or both. The isotope ^{59}Fe, for example, has a half-life of 45 days and decays by emission of a beta (β) particle to stable ^{59}Co. Naturally occurring ^{226}Ra is the isotope of radium (Ra) with mass number 226. Radium has 88 protons that characterize it, so ^{226}Ra has $226 - 88 = 138$ neutrons. It also happens to be highly radioactive.

The earth as it was formed included a variety of naturally radioactive materials. The 3 main chains of radioactive decay stem from the heavy elements uranium and thorium, but there are also a number of long half-life lighter elements that have remained through eons of time since the earth's formation or are constantly being formed by action of cosmic rays. The most important of these are carbon-14 (^{14}C), potassium-40 (^{40}K) and rubidium-87 (^{87}Rb). There is also a background level of tritium (^{3}H) produced by cosmic radiation, but more is produced from weapons tests and nuclear reactor operation.

It is worth following one of the uranium chains in detail because only by doing so can we understand the presence and behavior of 2 important radioactive elements—radium-226 (^{226}Ra) and radon-222 (^{222}Rn), the latter of which is a gas. The starting point of the chain is the nonfissionable isotope of uranium (^{238}U). It decays extremely slowly with a half-life of 4 billion years, giving off an α particle and producing thorium-234. The thorium-234 decays fairly rapidly with emission of a β particle to yield the element protactinium-234, which decays in turn quickly to uranium-234. This long half-life material (200,000 years) emits an α particle to become thorium 230, which in turn decays to radium-226.

We'll pause a moment here because you may feel this is a bit dull like the portion of Genesis in the Bible where Adam begat Seth and Seth begat Enos and Enos begat Cainan. Like the individuals in the Bible, some of the isotopes in the decay chain are more important than the others, and radium-226 is an important one worth a moment's digression. Radium-226 has a half-life of 1600 years and emits both α particles and γ rays. It was the earliest radioactive material to be used commercially and was specially separated from uranium-bearing carnotite ores to provide material to activate luminous paints and for medical radiation treatment. Remember that the unit of radioactivity, the curie (Ci), is defined as that amount of radioactivity having the same disintegration rate as 1 gram of radium-226.

Continuing our decay chain, the next step is radon-222 that results from α particle emission from ^{226}Ra. ^{222}Rn is a gas and is an α emitter with a half-life of 3.8 days. It thus has a high specific radiation intensity, meaning that a small quantity of the gas can give off intense radiation. We'll return to the significance of radon in a moment, but first let's follow the uranium decay chain to its conclusion. There are actually 7 more steps to the chain involving isotopes of polonium, lead, and bismuth. The final product is lead-206 (^{206}Pb), which is a stable, nonradioactive metal.

Thus our radioactive decay chain has gone from uranium of mass 238 to lead of mass 206 by way of 14 different radioactive species, each with distinctive and different radiation characteristics. It is not too difficult to see that the radon occurring with this chain, even though it has a short half-life and rapidly decays away, is always being replenished by the slow decay of radium-226 in turn replenished by the decay chain above it. Nature thus provides a constant and balanced amount of radon, always decaying but always being replenished as fast as it decays.

There are 2 other natural radioactive decay chains, 1 called the actinium series that starts with uranium-235 and ends with lead-207 and the other the thorium series that goes from thorium-232 to lead-208. Of these the latter is the more important, since it starts from a different element than the other 2 and thus generates radioactivity in areas where thorium is present and uranium is not. The thorium chain is dominant in major areas in India where thorium-bearing monazite sands exist.

Uranium and thorium are widely distributed about the world. In the United States the most concentrated deposits of uranium lie scattered through the mountain areas of New Mexico, Colorado, Montana, Utah, and South Dakota. Large total amounts of uranium exist, although in low concentrations, throughout the Southeast in the so-called Chattanooga shales, which contain 0.006–0.008% uranium, and throughout the Northeast in granite and other rock formations. Uranium is widely distributed in Florida and has been commercially produced there as a byproduct of mining operations for phosphates for fertilizer. The average abundance of uranium in the earth's crust is 2.8 parts per million (ppm), and the average soil near the surface contains 1.8 ppm. The figure in the earth's crust compares with 50,000 ppm for a common element like iron, 80 ppm for nickel, 70 ppm for copper, 16 ppm for lead, 0.5 ppm for mercury, 0.1 ppm for silver, and 0.005 ppm for gold. Uranium is thus commoner than a number of other familiar metals. Even seawater contains measurable amounts of uranium, so much so that the Japanese have a research program exploring the possibility of commercial separation of uranium from seawater. The radiation associated with the uranium decay chain is thus all-pervasive. This is particularly true of the radon component, since being a gas it can diffuse through ground and air.

It is also the radon component that gives us the most concern, not only because of its own radiation but because of the radiation from its decay products, often referred to as "daughter products." These daughter products are radioactive isotopes of bismuth, polonium, and lead. Radon, as we have mentioned, is a gas with a short half-life (3.8 days). It can move freely with the air in our rooms and with the air is breathed into and out of our lungs. Its decay products, though, are solids formed atom by atom from the radon decay. Wherever the radon gas penetrates it deposits the daughter products. They can attach to dust particles or settle out on surfaces. The last 3 members of the chain are lead-210, bismuth-210, and polonium-210, having half-lives of 19.4 years, 5 days, and 138 days respectively. They can thus remain for long times emitting β particles, α particles, and γ rays. Inevitably each of us has inside us now some small number of these radioactive atoms, most probably in our bronchial or lung tissue.

This is not a new phenomenon brought about by nuclear power. These materials have been with us since the beginning of time, and man and all his forebears have accommodated to their presence. The amounts are very, very small. The radon concentrations at ground level range from 10 to 1000 picocuries per cubic meter of air, and you will recall that a picocurie is one trillionth of a curie. Estimates have been made that radon dissolved in body tissue can give an average whole-body exposure of 3 mrem/yr, while estimates of lung dosage range from 50 to 450 mrem/yr. These are numbers worth remembering when we consider the added levels from artificially produced radiation.

Carbon-14 is another naturally occurring radioactive isotope of importance, since it is readily incorporated in biological organisms. ^{14}C has a half-life of 5730 years and emits a relatively weak β particle. It is formed in the upper atmosphere by interaction of the nitrogen in the atmosphere with neutrons from cosmic rays and has been present in constant amounts for thousands of years. Our bodies have the same fraction of ^{14}C as exists in the atmosphere, but this is an extremely small amount and produces only 16 disintegrations per minute per gram of carbon. Remember that a curie is defined as the amount of radioactivity that produces 3.7×10^{10} disintegrations per second, so the natural amount of ^{14}C in a gram of carbon is 7 picocuries. That is not very much. The average annual internal dose to an individual in the United States from ^{14}C is estimated to be about 1 mrem, however.

Living creatures constantly exchange their carbon atoms with those in the air or in food, so the amount of ^{14}C stays constant in spite of its slow rate of decay. After death this exchange no longer occurs, so in wood or very old cloth the decay of ^{14}C over thousands of years changes the ratio of ^{14}C to ^{12}C, the primary stable isotope of carbon. The change in this ratio is directly related to the age of the specimen or more exactly the period of years since it was living. This is used to determine the age of relics or archeological finds that are thousands of years old. The method is not sensitive beyond 40,000 to 50,000 years because by that time almost all the ^{14}C will have decayed.

Potassium-40 (^{40}K) is another significant natural radioactive isotope. It constitutes 0.012% of all potassium, and potassium is the seventh most common element on earth. ^{40}K has a half-life of 1.3 billion years, so it is probably still here from the original formation of the earth. The long half-life means a very slow rate of decay, but the decay takes two forms. The major one (88%) is emission of a 1.4 MeV β particle, but the minor mode (12%) gives off a 1.4 MeV γ ray. It is thus highly energetic even though not emitted in large quantity. Another significant fact about ^{40}K is that potassium is an essential ingredient in the growth of plants. ^{40}K is thus all-pervasive in the food chain in both vegetables and meat. The fertilizer industry helps spread potassium uniformly over all American farmland, as it properly should, but this assures a highly uniform distribution of ^{40}K also. The 5-10-5 lawn fertilizer you apply every spring and fall is detectably radioactive, since the last number of the three indicates that it contains 5% potassium, and 0.012% of that is potassium-40. This doesn't mean you should stop fertilizing your lawn, because the radiation levels are extremely low and man has lived with similar levels since the dawn of agriculture. You are probably receiving a yearly dose of about 17 mrems to your whole body from the ^{40}K it contains.

Rubidium is not a well known element because it has no major uses calling it to our attention. It is, however, the sixteenth most common element in the earth, commoner than zinc or nickel or copper. Nearly 28% of naturally occurring rubidium is the isotope ^{87}Rb, which is slightly radioactive (half-life 49 billion years) giving off a relatively weak β particle. A practical way of thinking about its radioactivity is that rubidium metal, if left on top of a photographic film, would expose it in 30 to 60 days. Rubidium is absorbed biologically, however, and contributes about 0.6 mrem whole-body dose to an average U.S. citizen in a year.

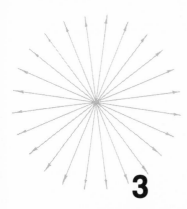

3

NUCLEAR REACTORS

We have now covered the origins and extent of the important naturally occurring radiation and can turn to the activities of man that add to our radiation environment. The one disturbing the natural balance most significantly and arousing great controversy is nuclear energy, particularly through the impact of nuclear weapons. In this chapter we will examine nuclear reactors, and in Chapters 4 and 5 cover the nuclear fuel cycle and nuclear weapons.

BASIC PRINCIPLES AND CHARACTERISTICS

It will first be necessary to establish some of the basic principles and characteristics of nuclear reactors. There is a unique jargon for nuclear power, and we will have to recognize and understand a series of new terms. We will start with fission, the fundamental reaction on which nuclear power depends, and then try to build from that the essential features of a nuclear reactor.

There is a unique characteristic of certain isotopes of the heavy elements uranium and plutonium that when struck by a neutron they will break up into two parts of unequal mass with simultaneous release of more neutrons and very large amounts of energy in the form of heat and gamma (γ) rays. This event is known as fission, and these isotopes of uranium and plutonium are categorized as fissionable. For our story about radiation it is important to know that the specific isotope fragments created by a fission event cannot be exactly predicted. They vary in random fashion, and for uranium-235, for example, we can predict only that on the average the fragments have the greatest probability of being either in the range of mass 85 to 105 or the range from 130 to 149. Some fragments are created with masses anywhere between mass 72 and mass 162, however. The fission products therefore consist of a very large variety of radioactive isotopes having half-lives from fractions of a second to thousands of years.

The main fissionable isotopes are uranium-235, plutonium-239, and uranium-233. Today's type of reactors depend mostly on ^{235}U, although there is some energy contribution from ^{239}Pu in them also. ^{233}U is not now important in nuclear reactors, although it could be if we elected an alternate design path. We will therefore discuss reactors for the moment in terms of ^{235}U.

Nuclear power exists because in each fission event induced by a neutron 2 to 3 more neutrons are produced. Each of these produces additional fissions and still more neutrons. Thus a chain reaction is created on the extraordinarily rapid time scale of nuclear events, and enormous amounts of energy can be released in extremely short times as in a nuclear bomb. To harness this energy we need a way to use up extra neutrons so that they cannot produce more fissions. It is not hard to see that if each fission event can produce only one additional fission there will be continuous production of power at a fixed level, whatever that level was when we achieved the 1 for 1 balance. If each fission produces on average more than 1 additional fission, power will grow; if less than 1, then power will decrease. We can exercise this control by insertion or withdrawal of poison materials such as boron, cadmium, indium, or hafnium, having the capability of absorbing huge numbers of neutrons in nuclear reactions that do not contribute significantly to the energy generated.

Once we can control the level of energy generation we also need to provide a coolant whereby the energy can be withdrawn and put to use. Coolants may be liquid or gaseous but must have the characteristic of absorbing very few neutrons in nonproductive or parasitic reactions. Water is an ideal coolant and can be converted to steam to drive turbines for electricity. However, regular water (H_2O) has a slightly higher absorption rate for neutrons than desirable, which leads to an interesting design choice in nuclear reactors.

Uranium as it comes out of the ground consists of 99.27% uranium-238 and only 0.73% uranium-235, the fissionable isotope. If we use this natural uranium in a reactor cooled by regular water, because of losses of neutrons by capture by the water and other poisons and because of losses of neutrons to the space beyond the reactor, we cannot sustain a chain reaction and produce power. One solution to the problem is to use uranium that has more ^{235}U than in natural uranium. We can obtain this by one of several isotope separation processes that produce uranium enriched in the ^{235}U isotope. These processes are complex and expensive but have been developed to a high degree of sophistication.

An alternate solution is to change the isotopic composition of the *water*. All water contains a very small amount (0.015% or 150 ppm) of deuterium, the hydrogen isotope of mass 2. We represent such water as D_2O instead of with the symbol H_2O used both for natural water and more specifically for water containing protium, the common hydrogen isotope with mass 1. D_2O absorbs fewer neutrons in parasitic reactions than does H_2O, and a reactor cooled with pure D_2O can sustain a chain reaction with natural uranium as fuel.

The point of this discussion is that these design considerations have led to two fundamentally different classes of water-cooled reactors. One class, called *light water* reactors, uses normal water (H_2O) coolant but requires uranium enriched to about 3% in uranium-235. The other class, called *heavy water* reactors, uses heavy water (D_2O) coolant but can be fueled with natural uranium. Light water reactors are commercial in the United States, Japan, and Europe, while heavy water reactors are commercial in Canada.

Before considering the nuclear reactor as a whole we must discuss neutron energy and the term "moderator." As emitted in a fission event neutrons have enormous energy, typically about 2 MeV. They are most effective in causing additional fissions, however, when they have been slowed down to much lower energies in balance with the thermal agitation of nuclei in that region. To achieve these thermal neutrons a material is needed that will "moderate" their energy (that is, slow the neutrons down) but without capturing many in the process. Light elements are best for this moderating process; water, because of its hydrogen content, is an excellent moderator. Carbon is also good.

We have now talked briefly about the four critical parts of a nuclear reactor: fuel, control materials, coolants, and moderators. It is time to consider how these go together in a practical nuclear power reactor and then how these reactors contribute to our radiation environment. For steam to do useful work such as driving a steam turbine to produce electricity, it must be at a relatively high temperature and pressure. It is not sufficient to generate steam by boiling water at atmospheric pressure. The kitchen pressure cooker is a way of getting water and steam to temperatures higher than 100°C of normal boiling, and this same principle is used in nuclear reactors. The lattice of fuel, which constitutes the nuclear core, is immersed in water in huge pressure vessels in which the water can be heated to about 300°C at pressures above 100 atmospheres. Pipes permit circulation of water or steam outside the pressure vessel to either a separate steam generator or to a steam turbine. Control rods containing boron or cadmium poison are inserted from either the top or the bottom of the pressure vessel and can be moved in or out of the fuel lattice to lower or raise power.

The nuclear fuel is in the form of stacked pellets of uranium dioxode (UO_2) ceramic, a dense, hard, black material, which is sheathed in an alloy of zirconium metal known as zircaloy. This protects the UO_2 from any harmful corrosion effects from the 300°C water. Depending on the particular reactor design, the pellets are from ⅜ to ½ inches in diameter and about ½ inch long. The tubes in which they are stacked are 12 to 14 feet long, and there are from 49 to 289 of these filled tubes or rods in an assembly. The type of reactor having 49 rods per assembly has 764 assemblies in a core for a total of over 37,000 rods, and another type with 204 rods per assembly has 193 assemblies in the core for a total of over 39,000 rods, so there will be typically from 37,000 to 40,000 fuel rods in the core of a large light water reactor producing in the order of 1000 megawatts of power. One megawatt equals a thousand kilowatts or a million watts, a lot of power in terms of a 100-watt light bulb.

In considering radiation from light water reactors we have to make an additional distinction between a boiling water reactor (BWR) and a pressurized water reactor (PWR). In a boiling water reactor, as the name implies, water is boiled directly in the reactor vessel. The core containing the fuel is always covered with water, but steam is generated at the surface of the fuel rods and passes up through the water to be collected in a space at the top of the reactor vessel and led through pipes directly to the steam turbine. The steam turbine drives an electric generator to produce the power fed to the utility network. It is important to recognize that in the BWR any radioactivity in the cooling water can be carried over with the steam into the turbine and into the turbine exhaust system. There are elaborate systems for trapping and filtering most of this radiation, but the emissions from a BWR run higher in general than those from a PWR.

In the pressurized water reactor (PWR) the reactor vessel is entirely full of water and at sufficiently high pressure to prevent boiling. The pressure in a PWR vessel is typically 2000 pounds per square inch (psi), while that in a BWR is only about 1000 psi. The high pressure hot water from the PWR vessel is pumped to separate steam generators. Here it passes through a very large number of small tubes, and its heat is used to boil water in a totally separate system surrounding the tubes at lower pressure. The significant point is that the steam produced does not contain any of the radioactive products from the reactor, at least as long as there are no leaks between the two distinct water systems. Thus the turbine exhaust from a PWR ideally should contain no radioactive elements. In practice the moderate frequency of small leaks between the primary and secondary systems does introduce some radioactivity in PWR exhausts. On average it is less than BWRs, however.

With this brief introduction to the nature of light water reactors we can begin to consider how they contribute to our radiation environment. We will look first at their contribution under normal operating conditions and then see what can happen under accident conditions.

LIGHT WATER REACTORS AND THE ENVIRONMENT

During normal operation of a light water reactor there are three paths that can generate radiation exposure to the public. The first path is through the radioactive gases that either occur as fission products or are generated by interactions between high-energy neutrons and the gaseous atoms in the air. The major fission product gases are xenon, krypton, and iodine in a variety of isotopes. The significant neutron activation products are nitrogen-13, nitrogen-16, nitrogen-17, fluorine-18, and oxygen-19. Although significant amounts of the oxygen and nitrogen isotopes are formed, they have half-lives in the order of seconds and decay away before they have any chance to escape from the reactor buildings. Only nitrogen-13 with a half-life of 10 minutes is ever emitted to the environment and that only at the rate of a few microcuries per second. The small amount and the short half-life make this isotope an insignificant factor.

Iodine can be readily absorbed in charcoal beds placed for this purpose in the gas cleanup system. Therefore very little gaseous iodine is emitted from either BWRs or PWRs in normal operation. Perhaps the largest routine emissions of iodine come from the turbine building exhausts in BWRs where occasional leakage from the turbines permits small amounts of radioiodine to enter the turbine room and be vented externally through the turbine ventilation system. The maximum observed release rate in a typical year is 7 to 8 curies of ^{131}I and almost all BWRs will be a factor of 10 to 100 below this. Converting even this worst case to human exposure says that an infant living at the plant boundary would receive 0.2 millirems (mrems) dose to the thyroid (where iodine concentrates) in a year. Remember that the average whole body background is about 100 mrem/yr. Radioiodine can also enter humans, particularly children, through the chain of grass to cows to milk to children. For a typical BWR a child drinking milk at the rate of a quart a day from a cow at the most critical distance from the plant would receive only about 1-mrem/yr thyroid dose.

The most significant gaseous radioisotopes emitted from a nuclear reactor are the noble gases xenon and krypton. The term noble means they are chemically nonreactive. This makes them harder to remove in the reactor gas cleanup systems, but also makes them less easily incorporated in humans. They are breathed in and breathed out and do not stay in the body. Their daughter products are not as insidious as those from naturally occurring radon, but they are a complicating factor for some of the isotopes. Krypton-85, which is the krypton isotope produced in most abundance, fortunately decays directly with a 10.7-year half-life to stable rubidium-85. The worst behaving krypton isotope is ^{89}Kr, having only a 3.15-minute half-life and decaying to strontium-89, having a 51-day half-life. ^{89}Kr is only significant close to plant boundaries, however, since in less than an hour it will all have decayed away. (Radioactivity levels are reduced by a factor of a thousand in 10 half-lives and by a factor of a million in 20 half-lives.) ^{89}Kr is in any event a relatively small part of the gaseous release from a reactor, perhaps 4 to 5 curies per year compared to 1000 curies for ^{85}Kr.

Six different radioisotopes of xenon gas are present as reactor fission products and like krypton can escape from the reactor fuel through occasional defects in the rod cladding. In BWRs such xenon can be carried with the steam into the turbine system and out through the process gas cleanup system and up and out through the 350 to 400 foot stack. These stacks are designed to create rising plumes of gas that blend in highly dilute fashion into the upper air masses. The actual dispersion pattern will depend on the weather: whether it is raining or sunny, the wind direction, the temperature, and many other factors. In PWRs the xenon can make its way out through the system via failed fuel plus leaks in the primary water system or leaks from primary to secondary system and hence through the turbine and out the turbine exhaust.

The most abundant reactor-produced xenon isotope is xenon-133. This has a 5-day half-life, but decays to nonradioactive cesium-133. It is thus only significant in its contribution to external whole-body radiation and to the lung. There is no lingering long-term radiation from daughter products. Xenon-135 is also produced in significant quantity and decays with a half-life of 9 hours to cesium-135, which is a beta (β) emitter of very long half-life (2 million years) and thus only weakly radioactive. Xenon-137 and xenon-138 are also produced in moderate quantity and have half-lives of 3.8 minutes and 14.2 minutes respectively. Thus they do not travel far from the reactor site and are not an especially significant factor, even though their daughters are radioactive cesium isotopes that deposit as solids. ^{137}Cs, the daughter of ^{137}Xe, is produced directly as a major fission product as well and is much more signficant as a portion of the liquid effluents.

You may be wondering by now what all these complex decays from a considerable variety of radioactive gases really mean, so we'll try to add it all up. We have shown that neither the activation gases (N, O, F) or gaseous iodine pose any conceivable problem in the vicinity of the reactor or beyond. The question of noble gases cannot be dismissed so lightly, however, and we must pursue it further. The output of radioactive xenon and krypton varies widely from reactor to reactor and depends considerably on the extent of fuel clad perforations in the reactor. These clad perforations permit water to contact hot fuel pellets directly, and although uranium oxide reacts only slowly with water under these conditions, some fission products will escape into the cooling water, and some part of them will eventually contribute to the radioactive effluent from the plant. There is thus a direct relationship between the number of fuel failures (i.e., clad perforations) and emission levels of radioactive noble gases.

These emission levels range from lows of a few curies per year to extreme cases of as much as a million curies per year. These sound like large numbers, but it must be remembered much of the radiation is from short half-life isotopes that decay away rapidly, and most of it is β radiation with very limited penetrating power.

The conversion from curies of radioactive gas released to millirem of exposure to an individual living close to the plant is a difficult one. It will depend on the height of the stack (if any) through which the gases are released, the typical weather patterns in the area, the mix of isotopes from that particular plant, the time the gases are held within the plant before release, etc. In the early 1970s the Nuclear Regulatory Commission (NRC) established that nuclear reactor design and operation objectives should be to limit exposure to any member of the public from airborne radioactivity to a maximum of 5 mrem/yr. In addition the NRC has emphasized the philosophy of ALARA, standing for "as low as reasonably achievable," as a means of minimizing radiation release from reactors and other sources as well. Typically a person at the boundary of a nuclear plant 100% of his time would receive from 1 to 10 mrems/yr as the plants are now operated. Both the National Council on Radiation Protection (NCRP) and the International Council on Radiation Protection (ICRP) recommend a dose limit of 500-mrem/yr whole-body exposure for any member of the public. Radiation levels from gaseous releases from normally operating nuclear reactors are clearly well below this recommended limit.

Effluent water is the second path through which reactor radiation can reach man. Many fission products remain dissolved in the coolant water and are circulated with it. When leaks occur, as they inevitably do, the leaked radioactive water must be disposed of. It is first collected in drains and sumps and then typically is filtered and degassed. Gaseous products are treated and vented as previously discussed, and the solids deposited in the filters go with the solid wastes. The filtered water then goes to a demineralizer to remove ionic species and finally to an evaporator. The liquid is then stored in large tanks and finally is filtered again and discharged via the clean liquid waste discharge. The effectiveness of these cleanup systems is such that the water eventually released has very low levels of radiation. Estimates have been made that swimming in such water for 100 hours per year would give a total whole-body dose of only 0.0001 to 0.00001 mrem/yr. Measured releases are in the range of a total of 0.1 to 10 curies per year diluted in very large volumes of water.

Typical long-lived fission products that can be present in tiny quantities are ^{89}Sr, ^{90}Sr, and ^{137}Cs. ^{58}Co and ^{60}Co may also be present from neutron activation of nickel in the structural materials of the reactor. It has been suggested that even these small amounts of radioactivity can be concentrated by preferential uptake by fish or shellfish that are then eaten by man. Calculation and measurement show that such concentration does occur, but the amounts stored are inconsequential, producing a dose of well under 0.01 mrem/yr even if an individual ate 50 grams of shellfish a day for the entire year. So even if there is a nuclear plant on your seacoast, enjoy your oysters without a qualm. Answering similar concerns for hunters, one study showed that if a wild duck obtained all his food within 200 feet of a reactor water outflow, and a hunter ate 8 pounds of such duck a year, he would get a maximum dose of 0.007 mrem/yr to his bone marrow, again a trivial effect.

Tritium, the isotope of hydrogen of mass 3, is also present in appreciable quantity in liquid wastes from reactors. This is not the only source of tritium in our environment, because some is generated by cosmic rays, some comes from fuel reprocessing, and there is a legacy of tritium from nuclear weapons tests. Reactors are a substantial source, however, and a large power reactor will release from 0.1 to as much as 1000 curies of tritium per year, mostly via liquid wastes. PWRs generate and release nearly 100 times as much tritium as BWRs. This is because in the early PWRs the uranium dioxide fuel pellets were clad with stainless steel instead of zircaloy, the zirconium alloy mentioned previously, and later units rely on boron in the cooling water to help reactor power control. The stainless steel permitted slow diffusion and escape of tritium generated in the nuclear fuel, while the high neutron flux in the reactor core generates tritium by absorption of neutrons in the boron. Tritium is also produced in both BWRs and PWRs in small quantities by interaction of neutrons with the small amount of deuterium (^{2}H) in the coolant water.

Tritium has a half-life of 12.3 years and thus remains as a radioactive entity in the atmosphere for a hundred years or more. On the positive side, however, it emits only a relatively weak β particle (18 keV), and although readily absorbed in the body replacing normal hydrogen, it is uniformly distributed in tissue and thus is biologically less damaging. It is also readily exchanged with normal hydrogen and has a *biological* half-life of 12 days. This means that of any given amount of tritium in the body, one-half will have exited in 12 days. It should not be confused with the *radioactive* half-life, which for tritium is 12.3 years. All things considered, tritium from reactors is not a significant health hazard.

The final way radiation can reach man in the course of normal reactor operation is through solid waste. Solids left behind on filters or from evaporation of radioactive liquids or used resins from water cleanup are collected, placed in drums, and shipped off-site for disposal by burial. Their hazard, if any, is transferred from the reactor site to the burial site. (See the section on waste disposal in Chapter 4.)

NUCLEAR REACTOR ACCIDENTS

To this point we have been talking about radiation generated by routine operation of nuclear power reactors and have seen that while radiation is introduced into the environment by such operation, the amounts are controlled within safe limits. The next obvious question is what about accidents. This is a very difficult question to consider objectively, because relatively few accidents have occurred to date, and we must talk in terms of what *could* happen rather than what *has* happened. The latter is a valid base point from which to consider the former, however.

Before describing real accident cases it is essential to get in perspective the fundamental safety concepts that exist in power reactors. Early reactor designers felt that serious accidents resulting in release of large amounts of radioactive fission products would be very rare events. Their philosophy was to build reactors in very isolated areas and let dilution and dispersion take care of any problems that could conceivably arise. The prospects of large numbers of reactors being built for electric power generation, however, led to the most important single safety concept in nuclear power: containment. The proponents of containment argued that accidents could happen to a reactor that would release dangerous amounts of fission products through fuel overheating and perhaps even melting. They urged that a sealed containment building be built around the reactor designed to withstand any likely pressure generated by such an accident and capable of retaining within its sealed domain all or almost all the escaped radioactivity. The opponents of containment argued this would add excessive and unnecessary cost and would seriously delay the advent of commercial nuclear power.

In this case safety won over economy, and all the U.S. and most foreign reactors have been built with sealed containment buildings surrounding the reactor pressure vessel. The concept proved its value many times over at Three Mile Island where the containment concept worked and the bulk of the large radioactive release was kept within the containment building. The Soviets disdained containment buildings in their early power reactors but are now using them in some designs.

A second key design feature of all reactors is redundancy. For every critical system there is a backup and frequently more than one. This defense in depth is an integral part of reactor design and is a major reason for the excellent safety record of nuclear reactors so far. No design can be completely foolproof, and our concerns must be with the probabilities of rare events and the severity of their consequences. We will first look at some of the most severe accidents that have happened so far and then consider the extent of future risk.

WINDSCALE. The most serious radiation release prior to Chernobyl occurred in 1957 at one of the Windscale reactors in England. These reactors are of a very different type than the pressurized water reactors we have previously described. They are essentially very large cubes of graphite through which pass horizontal fuel tubes filled with uranium metal fuel cooled with carbon dioxide gas. The graphite acts as the moderator to slow down the neutrons, a function accomplished by the water in light water reactors. In the intense neutron flux existing within the reactor, the graphite over a period of months or years slowly stores energy that can be released as heat relatively rapidly when triggered. Surprisingly enough the trigger is heat itself. When graphite containing this stored energy is heated high enough it spontaneously releases its stored energy and heats up even further. It was found that this stored energy could be safely released by very carefully warming the graphite by restricting reactor cooling. There would be a slow temperature rise while the stored energy was released, and then the graphite would return to its normal temperature free of its stored energy for many more months.

The procedure had been successfully used a number of times, but an attempt to use it in October 1957 resulted in serious trouble. All appeared to be going well with the main blowers off after the usual moderate temperature rise. The operator saw the temperatures start to drop and provided more nuclear heat to sustain the temperature for more thorough release of the stored energy. Unbeknownst to him, however, one region was hotter than normal, and the uranium in this region began to smolder and then to burn, eventually burning graphite and uranium fuel in a region of nearly 150 fuel channels. Operators who removed a fuel plug on the outside of the reactor could see glowing hot fuel inside.

They tried to cool the area down with the normal carbon dioxide gas coolant, but this failed, and they eventually had to flood the reactor with water to extinguish the burning. Radioactive emissions vented from the tall stack of the reactor for almost 24 hours from noon 10 October to noon 11 October, and substantial amounts of radioactivity escaped. It should be pointed out that the Windscale reactors had no containment building designed to contain the radiation in such an unexpected accident.

The amount of radiation released in this event is not known exactly, but reasonable estimates have indicated 16,000 to 20,000 curies of iodine-131, 16,000 curies of tellurium-132, 1100 curies of ruthenium-103, 600/1200 curies of cesium-137, 80 to 140 curies of strontium-89, and significant amounts of many other isotopes such as zirconium-95, ruthenium-106, and cerium-144. The wind was initially light and variable from the southwest and then became stronger from the north and west, blowing across the English Channel towards Denmark and Holland. The distribution of the radioactivity was thus both complex and international.

Measurements directly under the plume of the stack about a mile downwind shortly after the accident showed direct γ exposure of 4 milliroentgens (mR) per hour. In this same region a total exposure of 10 to 20 mR would have been received in the following week from γ radiation from particles deposited on the ground. This was probably the area of maximum dose, so total exposure of people at this point of highest exposure was still not significant relative to the normal 100 mrem of exposure from natural sources in a year. The question of inhalation and intake via contaminated food was somewhat more serious, however.

It was clear to public health officials that the isotope of greatest concern was ^{131}I since it was emitted in the greatest quantities, could disperse as a gas, and concentrates in the thyroid when it is ingested into the body. Unlike light water reactors, Windscale had no charcoal filters to trap out iodine, and about 20,000 curies of it was thus freely emitted into an area where dairy farming was a major occupation. Measurements after the accident showed that contamination was occurring in the milk of cows grazing over a considerable area around the plant. Measured amounts increased from a trace to as high as 0.4 to 0.8 microcuries per liter (μCi/l). A decision was therefore made to restrict the distribution of milk from an area of about 200 square miles. The distribution restriction remained in force until the measurements dropped to 0.1μCi/l. This was achieved in most regions within 25 days, but in the most severely contaminated region milk distribution was not resumed until 44 days after the accident.

There has been extensive critical evaluation as to whether the radiation levels from the Windscale accident caused any serious harm to the health of people in the region. The answer is not very clear, but a case can be made that a slight increase in the number of cases of thyroid cancer might be expected in the region over the period of the lifetime of the people living near the reactor at the time of the accident. One apparently objective study estimates, for example, that over that lifetime period an increase of 260 cases of thyroid cancer might be expected, of which 13 might be fatal. This would be at a rate of about 6 to 7 cases a year and must be considered in light of the normal annual rate of new thyroid cancer cases in England and Wales of 600 to 700. It is doubtful that these statistical differences, if they really occur, will be large enough to be seen in the medical records of the region.

The Windscale accident was unique in the amount of radioactivity emitted over a large area. Such health consequences as may have existed stem almost entirely from the iodine-131 released, and as we will see later the likelihood of similar releases from water-cooled reactors is extremely small. Windscale, however, was a much more serious accident than Three Mile Island, which we will discuss later in this chapter. While the damage to human health was relatively small if it exists at all, there was economic harm to the region through curtailment of milk production, and the Windscale reactor itself was so severely damaged that it was permanently shut down.

SL-1. One of the very few reactor accidents involving immediate fatalities occurred 3 January 1961 at the SL-1 experimental natural circulation boiling water reactor at Idaho Falls, Idaho. This was a small reactor designed to produce 200 kilowatts of electrical power and 400 kilowatts of space heat for remote military installations. Being small and at a remote area away from any significant population, the reactor had been built within a cylindrical building made of ¼-inch-thick steel plate, but with conventional doors and not designed as a containment building.

The reactor had been shut down on 23 December for maintenance and modification. Previous experimental operation had resulted in adjustment of all the control rods in such a way that almost all the control function was concentrated in a single center rod of the five normally used. On 3 January a 3-man crew was assigned to a late shift to reassemble the control rod drives and prepare for start-up. At about 9:00 P.M. an alarm sounded in the nearby fire station. The firemen arriving at the reactor found no fire, but found very high radiation levels and were unable to get a response from the crew. Readings of 25 R/hr were measured in a reactor support building adjacent to the reactor proper, and cautious readings through a door of the reactor building showed 1000 R/hr inside. Remember that 400 to 500 R is a fatal dose to half of those receiving it. Thus the men inside for the 15 minutes or so since the incident that triggered the alarm had received an enormous dose of γ radiation. Two of the crew men were found on the floor of the reactor building, one already dead and the other dying. The latter was removed by 11:00 P.M., but was pronounced dead shortly after. The body of the third man was found pinned to the ceiling structure. It is believed that he had been manually adjusting the critical central control rod and had somehow removed sufficient control to permit an enormously rapid power escalation. The pressure generated had blown apart the reactor core and hurled the control rod up to the ceiling carrying the crew man up with it.

In spite of the force of this tragic accident almost all the radioactivity released from the reactor core stayed within the building. Total iodine release was estimated at 4 curies (compared to 20,000 at Windscale), and even without a containment building readings at the security fence 200 feet from the reactor building were 250 mR/hr dropping to 2 mR/hr 2000 feet from the boundary. Sudden death is always grim, yet in this case there were no medical consequences beyond the reactor building itself.

BROWN'S FERRY FIRE. The Tennessee Valley Authority's Brown's Ferry site was the scene of perhaps the most unusual nuclear accident. Units 1 and 2, both 1000-megawatt BWRs, were operating 22 March 1975 at full power. There had recently been some maintenance to the control cables leading from the reactor to the control room. These cables must pass from a cable spreading room through the heavy concrete walls protecting the control room. This is accomplished by a series of 10 cable trays, flat penetrations through the concrete. The trays are sealed with a foam-type polyurethane insulation, which on the day in question had just been replaced. The pressure in the control room is maintained slightly higher than in the reactor containment so that air flow, if any, would always be from the control room to the reactor; there could be no opportunity for radiation from contaminants in the air in the reactor building to leak back into the control room. Inside the control room a maintenance man wished to see if any air leaks existed in the newly installed sealant. He therefore held a lighted candle near the freshly sealed area, expecting that the draft created by any leak would make the candle flame waver towards the leak. His expectations were more than fulfilled. There *was* a leak, and the candle did more than waver. The flame was sucked into the leak where it ignited the polyurethane sealant. The sealant smoldered for a while, and the fire slowly spread and ignited cable insulation as well. All this was occurring in the sealed duct behind the control room wall, and the fire was not noticed for a half-hour, by which time it had spread significantly. There was concern that the use of water to put out the fire would cause electrical short circuits, so carbon dioxide gas was used to try to douse the fire. It did not succeed.

As soon as the fire was noticed, the reactors were shut down, but even under shutdown conditions coolant water is essential to remove radioactive decay heat. Because of fire damage to the control cables, normal control functions were being lost one by one. Enough backup systems and manual operation modes existed, however, so that coolant level was maintained at least 4 feet above the reactor core at all times. Finally, nearly 6 hours after the fire was first noticed, the decision was made to flood the cable areas with water. The fire was soon out and no additional problems were caused. The fire had begun shortly after noon on 22 March and was out by 6:50 P.M. By 9:30 A.M. the next morning the plant was in normal cold shutdown and all was secure.

In spite of massive damage to the control wiring, no damage was caused to the reactor core, no one was hurt, and no radiation was released. There were large economic losses to TVA to replace all the damaged cables and pay for replacement power during the long shutdown that followed, but the "defense in depth" philosophy on which the BWR design was based had worked, and there was no hazard to the public. The defense in depth philosophy would be more sorely tested 4 years later at Three Mile Island, but here again, when all was said and done, it worked too. We will look at that now in more detail.

THREE MILE ISLAND. Three Mile Island is the site of a twin unit PWR station near Harrisburg, Pennsylvania, operated at the time of the accident by Metropolitan Edison Company, a subsidiary of General Public Utilities. In the early hours of 28 March 1979 maintenance workers on Unit No. 2 accidentally caused the shutdown of a pump that was part of the circulation system returning condensed steam from the turbine to the steam generator. In such circumstances the design calls for automatic shutdown of the main feedwater pump to prevent damaging it, and an auxiliary feedwater supply to the steam generator is started up. There was a problem here, however, because valves in the auxiliary line that should have been open were closed; so for 8 minutes no coolant water was entering the steam generator. Remember that in a PWR hot water from the reactor boils the coolant water in this secondary system, but at the same time the presence of this cooling water helps maintain the temperature and pressure in the primary reactor loop.

Since there was no water in this secondary side of the steam generator, the pressure began to rise in the primary reactor loop, and, in accordance with design, a pressure relief valve opened and the reactor control rods were inserted, shutting down the reactor. A shut down reactor, however, continues to generate heat from residual radioactivity in the fuel. Right after shutdown this is about 5% of the heat at full power, so cooling of the fuel is still required. So we now had a reactor in which no more fission heat was being generated but substantial radioactive decay heat was still present. The open relief valve dropped the system pressure as it should, but at this point one of the critical events of the accident occurred. The relief valve did *not* reclose when normal pressure was reached, but the control panel indicator light told the operator that it *had* closed. The operators thus did not realize that coolant was still being lost.

For 2 hours and 22 minutes this valve remained open permitting coolant steam and water to escape, even though a second line of defense existed in the form of a blocking valve that could have been closed had the operators realized so much coolant was being lost. This nonrecognition of the extent of cooling water loss led to other operator actions restricting the coolant flow when it should have been increased instead. The reactor core became partially uncovered allowing extreme overheating of the core and severe damage to the fuel elements, associated with major release of fission products into the coolant water and steam. Some of the escaped radioactive water was transferred out of the reactor containment building into the adjacent auxiliary building where it overflowed a storage tank and permitted escape of radioactive noble gases into the auxiliary building and into the atmosphere outside. This was a breach in the containment system, but it was not a major one, and on the whole the containment system did its job and prevented really serious large releases of radioactivity.

It is now estimated that 10% of the total noble gases in the core escaped into the atmosphere. This would amount to about 10 million curies of xenon and krypton or about 10 times the maximum amount emitted in a year by any reactor operating normally but with some failed fuel. Release of iodine isotopes was very small (15 to 24 curies) since most of the radioactive iodine was absorbed in the water and remained in the plant or was absorbed in the charcoal filters in the gas cleanup system. Essentially all nonvolatile fission products, the bulk of the radioactivity, remained in the fuel, in the reactor coolant water, or within the containment building.

Both the Kemeny Commission appointed by the president to investigate the accident and the Rogovin Commission appointed similarly by the Nuclear Regulatory Commission agree that the radiation releases were too small to produce measurable harm to the surrounding population. The monumental cleanup task is still continuing in 1986, however. The decontamination process requires disposal of large amounts of radioactive waste, mostly in solid form since the radioactive wastewater is being treated to remove all contaminants and retain them in filters and resin beds. The safety record in the cleanup has been good, but the legacy of fear and mistrust, heightened by overly sensational and sometimes less than accurate reporting of the accident when it occurred, still resides in the neighborhood.

WHAT MIGHT BE

The strong objection to nuclear power voiced by many people and expressed in marches, picketing, sit-ins, legal obstruction, and attempts at legislative curtailment stems not from what has occurred, but what might occur. We must, therefore, look at what could happen, its probability, and consequences. The most exhaustive study of this subject was carried out at the Massachusetts Institute of Technology between 1972 and 1975 under the direction of Norman Rasmussen. It involved 60 people, 70 man-years of human effort, and cost 4 million dollars. It has been both praised and damned, but it still stands as the most rigorous scientific attempt to quantify the risks associated with nuclear power reactors.

The Rasmussen report makes it perfectly clear that the danger in an accident at a nuclear power plant comes from escaped radioactivity and not from explosive force as in a nuclear bomb. While we have seen from the SL-1 accident that very rapid power increases in a nuclear reactor can develop explosive forces sufficient to badly damage the reactor internals and drive mechanical parts out of the reactor, these are explosions caused by rapid thermal generation of steam and are not true nuclear explosions such as occur in nuclear weapons. They are thus far less formidable and pose no danger to people outside the plant except as they permit the escape of radioactivity.

It is thus radioactivity release with which we must be concerned, and we need to consider what the maximum amount could be, what is the probability of its release, and what the consequences would be if it were released. We have pointed out that radioactive fission fragments are generated continuously in the fuel as the reactor operates, and the longer the fuel has been operating the greater is the amount of these fission products. We have also noted that it is the gaseous or volatile fission products that are the problems under accident conditions. Rasmussen emphasizes that actual melting of the fuel is the only way in which really large amounts of radioactivity can be released. The Three Mile Island incident has shown that fuel disintegration short of melting but induced by reaction with steam at very high temperatures can also release substantial quantities of radioactivity. In either case the problem can only arise when the reactor core ceases to receive adequate cooling. How can this happen, and how likely is it?

One way it could happen is by mechanical failure and operator error such as occurred at Three Mile Island. Here a considerable portion of the core was uncovered and received no water cooling for perhaps as much as an hour or more. In such a circumstance the fuel is not completely dry or uncooled, because steam exists around it in a chaotic regime probably including some water carried in droplet form as well. Even so it is now believed that as much as 5 to 25% of the fuel in the reactor did melt and "slump" toward the lower part of the reactor pressure vessel where it recongealed. Much additional fuel disintegrated in coarse powder or fragments. The fact that actual melting occurred is important, because it belies the commonly held belief that core meltdown would result in a fiery ball that would melt its way through the pressure vessel and eventually breach the containment and release massive amounts of radioactivity. The term "China Syndrome" used for the title of a 1979 Jane Fonda/Jack Lemmon movie refers to this concept of the molten ball of fuel eating its way through the earth to China.

It is a nice dramatic concept, but the world doesn't work that way. If fuel melting does occur, any supercritical consolidation of molten material would quickly blow itself apart in minor explosions that would spread material about rather than concentrate it. Three Mile Island shows that even when there is grossly inadequate cooling both melting and nonmelting modes of fuel disintegration can occur without catastrophic damage to the pressure vessel or the containment. The probability of melting is less than the probability of other forms of disintegration, and the probability of containment breach in either mode is very small indeed.

It is clear that on rare occasions serious malfunctions or accidents can occur that will cause melting or disruption of the nuclear fuel and release substantial quantities of radioactivity. In most cases the containment system will function to retain the radioactivity, even in the case of severe core melting. Nevertheless we must still consider the consequences of a situation involving both serious core damage and a major containment breach. This was the spectre that made authorities consider evacuation of the area around Three Mile Island and is the main concern of those opposed to nuclear power. The consequences would vary depending on the weather at the time, particularly wind velocity and direction and whether it was clear, cloudy, or raining. They would vary depending on the reactor type, how long it had been operating, and the nature of the damage.

The Rasmussen-MIT study considered a great variety of combinations from the most likely to cases where all the worst possibilities combined in unlikely ways to create the worst possible accident. Computer codes were developed to calculate the probabilities of such events and their consequences in radiation release and harm to humans and property. The conclusions of the study have been challenged by parts of the nuclear industry as being too pessimistic and by antinuclear groups as being the opposite. It is fair to say, however, that the study has attempted to develop a logical approach to putting numbers to an area of prediction previously dominated by propaganda and guesswork. The study is not perfect, but it is rational and has provided us the best guide there is.

It is difficult to distill a complex analysis of events, probabilities, and consequences into a single essence. Some overall conclusions, however, can provide perspective on the whole issue of reactor hazard. The worst possible accident that could be conceived in the Rasmussen study was one involving total loss of cooling, failure of emergency core cooling systems, core melt, and breaching of both the pressure vessel and the containment building. The probability of such an event occurring was calculated to be 1 in a billion per year of reactor operation, but if it occurred it could result in 3300 early fatalities, 45,000 early illnesses, and 14 billion dollars of property damage. It is these numbers which, in spite of the exceedingly low probability of their ever happening, have been used to show that nuclear power is too dangerous to be allowed in populous areas.

The Rasmussen study was made before the accident at Three Mile Island. When this accident was matched against the Rasmussen analysis it was found that considering the seriousness of the damage and the partial breaching of the containment system, the predicted radioactivity release would have been much larger and the effects more severe than actually occurred. Why had this been? A special committee was established by the American Nuclear Society, with the collaboration of the U.S. Nuclear Regulatory Commission, the U.S. Department of Energy, and the Electric Power Research Institute, to examine this question and see whether there were fundamental reasons that would bear on the whole issue of reactor safety. The report of this "Special Committee on Source Terms" was published in September 1984 and contained the interesting conclusion that the "source term" as used in the Rasmussen study was too large by from 1 to several orders of magnitude; that is, it was too large by a factor of 10 to 1000.

The source term defines the amount and type of radioactive materials "available for escape to the environment from a reactor plant which has undergone a severe reactor accident." In considering the release of radioactive iodine, for example, the Rasmussen study had assumed that iodine would be present largely as a gas, and that in some cases as much as 90% of it would escape from the reactor. From a detailed analysis of the Three Mile Island accident and all other reactor accidents plus a careful review of the chemistry involved, the source term study concluded that almost all of the iodine would in fact be present as cesium iodide, a much less volatile compound. This cesium iodide would form tiny droplets, even at high temperatures, which would dissolve in any water available or deposit out on surfaces rather than escape out into the outside air.

Since ^{131}I would be one of the main dangers to humans downwind from a reactor accident, this finding is very important in estimating what the risks are from the maximum possible accident. Where the Rasmussen study indicated the possibility of as many as 3300 deaths from such a disaster, the new studies indicate this estimate to be much too high and that the maximum number of deaths, even in such a low probability event, would be anywhere from 0 to a maximum of 330.

Can we convert abstract numbers such as this into something meaningful in every day terms? One way is to think about what this means in a world where there are 1000 operating nuclear reactors (as compared to about 100 in the United States today). These numbers tell us that in such a world there would probably be an accident of that magnitude once every million years. In the past 100 years the greatest single man-made accident was the methyl isocyanate gas leak on 3 December 1984 in Bhopal, India, where it is estimated that over 2000 were killed and 200,000 made ill. Even those who apparently recovered may suffer life-shortening side effects. The failure of a dam in Johnstown, Pennsylvania, in 1889 cost about 2000 lives; the explosion of a French munitions ship in the harbor at Halifax, Nova Scotia, in 1917 killed over 1600 people; the sinking of the Titanic in 1912 cost 1500 lives; and automobiles kill about 55,000 people every year in the United States alone. So this one-in-a-million maximum nuclear accident is smaller in magnitude than other man-made accidents in our recent experience insofar as immediate effects are concerned. As in the Bhopal disaster there would be additional delayed effects, but in the worst possible Rasmussen scenario, reduced by a factor of 10 for source term correction, additional cancer deaths per year in the period 10 to 40 years after the accident might be, but probably would not be, as great as 330. This would be in a population of about 10 million people who might receive some radiation exposure beyond normal. This same population of 10 million people would normally have 17,000 cancer fatalities per year, so under the worst conceivable circumstances the accident would increase the cancer fatality rate by only about 2%.

For additional perspective we need to look at the magnitude of natural disasters over which man has no control. Six thousand people died in the Galveston hurricane in 1900, and in the period 1900–1972,

12,577 deaths in total were attributed to hurricanes. The San Francisco earthquake killed about 750 people in 1906, falls kill 18,000 people per year, and there is always the possibility of a major meteorite striking a large city. One analysis calculates that based on known meterorite impacts in the past there is a 1 in 10 million chance of a meteorite strike in the United States killing more than 10,000 people.

Adding this all up we conclude that the consequences of a worst possible nuclear accident, while severe, are less than those of other natural or man-made catastrophes that have been or will be experienced. We also conclude that the likelihood of having an accident of this maximum severity is very, very small. Compared to the probability and the consequences of war, for example, the probability/consequence summation for nuclear reactors is insignificant. We need to focus all our technical expertise on ensuring the safety of nuclear plants, but we do not need to abandon the benefits of nuclear power because of the risks inherent in the reactors themselves.

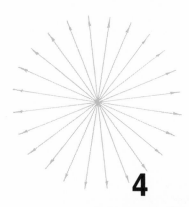

4

NUCLEAR FUEL CYCLE

THE OPERATION OF NUCLEAR REACTORS requires first the preparation of the nuclear fuel and finally the disposal of nuclear wastes. These operations contribute to the radiation environment of the world, and we need to consider their impact too. We can simplify this task by separating it into a fuel preparation and a waste disposal portion.

FUEL PREPARATION

As we have mentioned, uranium exists in many places all over the world, but the major sources are in Czechoslovakia, Australia, South Africa, Canada, and the United States. Most U.S. mining is in New Mexico and Wyoming plus lesser amounts in Utah, Colorado, and South Dakota. The potential exists also for uranium recovery in Florida as a by-product of the phosphate fertilizer industry.

The mining of uranium is basically similar to the mining of any other metal or the mining of coal. Uranium can be obtained from underground mines or from surface stripping. It is not found in very high concentrations, however, and a typical ore may contain only 0.1 to 0.3% uranium as uranium oxide. Thus large amounts of rock must be removed and treated in order to obtain and separate out the uranium. Because of this the preliminary mechanical and chemical separation is often carried out close to the mine site, and concentrates are shipped for further refining. In consequence the "milling" sites where this separation is accomplished must dispose of large quantities of residue or "tailings." These tailings are largely crushed rock or sand but also contain a residual amount of uranium plus many of the radioactive by-products such as radium, thorium, and other members of the uranium decay chain. It is

49

important to recognize that the mining and separation doesn't *add* any more of these materials to the environment, it merely brings them up from underground and collects them in a single area on top.

The total amount of radiation contained is not very large. A ton of typical uranium ore will contain about 7200 microcuries (μCi) of radiation, and about 70% of this or 5000 μCi remains in a ton of tailings. All the tailings generated in the United States up to 1970 amounted to about 80 million long tons[1] and contained a total of about 400,000 Ci of radioactivity. By comparison a single nuclear reactor after a normal period of operation would have an inventory of 8 billion curies.

In terms of world contamination, therefore, the mining and milling operations don't generate any new radiation, they merely redistribute it in undesirable ways. The problem is how to return it to the more benign locations typical of its origin. This can be accomplished by burial at modest depths, since about 20 feet of earth cover will reduce the radon emanation to natural soil values, and it is the radon activity that is the dangerous component. It is unfortunate that the seriousness of radon emanation from mill tailings was not recognized in the early mining efforts. In some cases the tailings were merely leveled with little or no earth cover, and there have been instances where contractors have used mill tailings as fill for house foundations, resulting in undesirable exposure to the occupants. A Uranium Mill Tailings Radiation Control Act was passed in 1978 giving the Environmental Protection Agency authority to issue standards for mill tailing disposal. The same act authorized the Department of Energy to take remedial action at inactive sites. The mill tailings situation should therefore be under adequate control in the future. There has also been concern that the dust created in milling operations can cause radiation problems in the vicinity of the mill. This is a valid concern in areas within about 2 miles downwind of an active mill, where radon readings may be significantly above background and there will be some deposition of fine particles of ^{238}U, ^{226}Ra, and ^{210}Po in particular. The amounts of particulates are quite small (typically perhaps 0.01 Ci/yr), and the mills are located in sparsely settled areas so the overall effect of the particles is small. This is not much comfort, however, to the rare individual who does live close to a uranium mill. Radon levels close in constitute the most significant hazard.

There have been cases of slight contamination of streams adjacent to uranium milling sites. Control via holding ponds and filtration is fairly simple, however, and is now pretty uniformly in place.

---

[1] A long ton is 1000 kilograms or about 2200 pounds, compared to our standard ton of 2000 pounds.

CONVERSION

The next step in the nuclear fuel cycle is conversion of the uranium concentrate to a gaseous compound named uranium hexafluoride (UF_6). This occurs because of the need to "enrich" the uranium in the ^{235}U isotope. We mentioned earlier that it is the ^{235}U which is fissionable in a nuclear reactor and that the light water reactors require 2 to 4% ^{235}U in their fuel in order to operate. Since the uranium as mined contains only 0.7% ^{235}U, the level of this isotope must be increased. This is accomplished by passing the UF_6 gas through a succession of porous barriers. The light isotope (^{235}U) can diffuse through the barrier slightly more easily than the heavier ^{238}U, and the percentage of ^{235}U is very gradually increased as the gas passes through successive barrier stages. About 1700 stages are necessary to enrich from 0.7% to 4.0% ^{235}U. The only radiation hazard from operation of a gaseous diffusion plant is from leakage of UF_6 gas. This has been kept very small through careful technology aided by the fact that the gas turns into a solid below 56°C (133°F) and escape from the plant is thus unlikely.

New isotope separation plants will use gas centrifuge technology instead of gaseous diffusion. In this technique very high speed centrifuges spin the UF_6 gas, and the heavier ^{238}U atoms tend to move to the outside compared to the lighter ^{235}U atoms. Again large numbers of successive stages are required, but the overall economics are more favorable. As far as radioactive release is concerned the two techniques should be comparable.

In either case the enriched UF_6 is transported in pressurized cylinders to a fuel fabrication plant where it is converted to uranium dioxide and made into pellets for insertion into the zirconium tubes of the reactor fuel assemblies. In January 1986 there was an accident involving the rupture of a 14-ton tank of UF_6 gas at the Sequoyah Fuels, Inc. plant in Gore, Oklahoma. One worker was killed and there was considerable press publicity about the accident mostly emphasizing the radioactivity involved. While UF_6 is mildly radioactive because of the contained uranium, it is hazardous largely because of its chemical activity, not its radioactivity. When in contact with moist air much of the UF_6 is converted to toxic hydrofluoric acid (HF), and it was the chemical toxicity of the UF_6/HF combination that killed the worker, not the radioactivity. It is true that liquid waste effluents from a fuel fabrication plant will contain significant amounts of ^{226}Ra, ^{234}Th, ^{230}Th, ^{234}Pa, which must be disposed of in a safe manner. Their total amount annually in the United States is only 1 to 3 curies, however.

NUCLEAR WASTE DISPOSAL

Since we have discussed in a previous section the radiation produced by nuclear reactors themselves, we can now consider one of the most controversial areas in nuclear power—the processing and disposal of the spent nuclear fuel after it leaves the reactor. Unlike the chain from uranium mining to final fabrication, a nuclear reactor enormously increases the amount of radiation in the world. We mentioned before that a single 1000 megawatt reactor will generate an inventory of 8 billion curies of radiation within the fuel before it is discharged. This is the radioactivity associated with the great variety of fission products produced from the fissioning uranium and plutonium and also includes the isotopes of plutonium, americium, and curium. The latter two elements do not exist in nature but are generated in the reactor through the capture and retention of neutrons. They are included in the class of elements referred to as transuranics because they are beyond uranium in the periodic table of the elements. They are mostly alpha (α) emitters, but have long half-lives (from 163 days for ^{242}Cm to 24,400 years for ^{239}Pu).

As long as these fission products and transuranics are contained within the reactor they pose no problem. We have already discussed the risks of their escape through accidents, and we must now consider how to dispose of them when the life of the fuel is exhausted. There are two alternate paths that may be followed. One is to chemically reprocess the fuel and recover the unused uranium and the generated plutonium. The other is simply to remove the fuel elements, place them in shielded casks, and dispose of them in either controlled repositories or deep burial sites. We will discuss first the reprocessing and recovery and then consider the alternates.

In the early development of nuclear power it seemed obvious that chemical recovery was the course to follow. A spent-fuel element contains in the order of 1% ^{235}U plus almost as much fissionable plutonium. If purified and reclaimed, both can be recycled back into light water reactors, or the plutonium can be used as fuel in breeder reactors.

We'll have to digress for a moment here to explain the fascinating concept of the breeder reactor, for it is essential to the understanding of the value of reprocessing. To understand the breeder you must first be reminded that uranium as it occurs naturally is 99.3% ^{238}U and 0.7% ^{235}U, of which only the latter is fissionable and therefore usable as fuel. Thus on first look it would seem that only a very small part of the uranium in the world could be converted to energy. It so happens, however, that some of the neutrons generated by fission of ^{235}U are captured by ^{238}U and converted via 2 intermediate steps to ^{239}Pu, which is fissionable. Thus nonfissionable ^{238}U becomes fissionable ^{239}Pu, which can be used as a fuel. Now comes the truly remarkable clincher. If ^{239}Pu is used in a special type of reactor, particularly one cooled with liquid sodium and without moderators to slow the neutrons down, and if ^{238}U is incorporated in the reactor also, the fissioning plutonium not only provides heat energy, but it provides enough neutrons to generate more plutonium from ^{238}U than is used up in the process.

Let's suppose, for example, that we had a reactor that over a period of time burned 100 units of plutonium but converted 110 units of uranium-238 to more plutonium at the same time. At the end we would have enough plutonium (100 units) to start another cycle and 10 extra to divert to another reactor. Thus in time we could convert all the ^{238}U to energy via the plutonium route. The breeder, therefore, permits us to generate perhaps 100 times as much energy from the world's uranium as could be generated without it. This vista cannot be opened without the breeder, and the breeder is not possible without reprocessing to provide plutonium.

This brings us back to reprocessing with the new recognition that in addition to concentrating the wastes for final disposal, it is also a vital step in providing a unique and major increase in our energy supply. We must see, however, how the reprocessing itself contributes to our radiation environment and then consider the problem of the final disposal of the concentrated wastes.

The basic steps in reprocessing are first a mechanical chopping of the fuel elements into small pieces, followed by chemical dissolving in concentrated acid. The acid solution is then treated to separate the uranium and plutonium from the remaining radioactive wastes. These very highly radioactive wastes can be kept in solution as liquids, or the liquids can be evaporated and converted into glasses or other stable solids.

While the chemistry of the process is not inordinately complicated, the extremely high radiation levels in the waste products introduce major engineering problems. All the mechanical and chemical operations must be carried out in huge concrete canyons beyond walls many feet thick. Viewing is through equally thick special glass windows or via shielded binocular optical systems. Mechanical manipulators are controlled on the safe side of the wall to carry out delicate and precise maneuvers in the fiercely radioactive interior of the canyon. When problems occur in the operation, as they sometimes do, they must be fixed by remote control devices or the radioactive material must be drained or removed to a storage section and sufficient cleanup must be carried out to permit brief entry of suitably protected repairmen.

A reprocessing plant normally generates four basic products: plutonium, recovered uranium, low-level wastes, and high-level wastes. The plutonium is suitable for use in either nuclear power reactors or in weapons, and the uranium can be returned to the fuel cycle. The low- and high-level wastes must be suitably disposed of. In addition, the chemical dissolving of the solid nuclear fuel results in the release of gaseous and volatile radioactive fission products. The gases from the dissolving tanks are treated by scrubbing with water sprays and by filtration. This removes most of the iodine, but very large amounts of ^{85}Kr (krypton) and ^{3}H (tritium) escape. The amount of tritium will vary with the specific reprocessing plant and its method of cleanup, but all the ^{85}Kr contained in the fuel will be released to the environment.

The amount of ^{85}Kr radioactivity in spent fuel is about 11,200 Ci/ton. Putting it another way a 1000 megawatt nuclear power plant (a typical large plant) will generate 375,000 Ci of ^{85}Kr/yr. This isotope has a half-life of 10.76 years, so reprocessing will continue to increase the total amount of ^{85}Kr in the world environment. On the other hand ^{85}Kr is an inert gas and decays by β particle emission, although 0.4% of the decays have a 0.5 MeV gamma (γ) associated with them. In all cases the final product is stable rubidium. ^{85}Kr is therefore a relatively harmless radioactive species with high human tolerance to it. There was great public outcry against the release of 50,000 curies of ^{85}Kr from the crippled Three Mile Island reactor. Yet the European reprocessing plants at Le Hague in France and Windscale in England have quietly averaged a daily emission of 4700 Ci ^{85}Kr between them every day since the TMI accident. Are these amounts hazardous? All medical evidence says they are not, since the krypton is not retained in the body, the radiation is weak, and the daughter product is biologically harmless. It is clearly wise to ensure adequate dispersal and dilution. We must also monitor and control carefully the total ^{85}Kr buildup in the world. Calculations show that with continued buildup of nuclear power and no additional control, the average concentration of ^{85}Kr in the air in the northern hemisphere could reach 1 nanocurie per cubic meter (nCi/m^3) by the year 2000. At this level each person in the United States would receive a whole-body exposure of 0.04 mrem each year in addition to the 100 mrem or more he or she would normally get. Biological effects at that level would be far less than those from the 0.2 to 0.3 nCi ^{222}Rn/m^3 that has always existed from natural sources. The amount is not entirely trivial, however, and the Environmental Protection Agency has recently issued new standards limiting release after 1983 to 50,000 Ci/1000 MWyr (megawatt years) of electric power generation. This will cut the total emissions resulting from the operation of a typical large reactor by a factor of 7 to 8.

The tritium released in reprocessing is a second area of concern. Total tritium generated in a nuclear reactor is about 20,000 Ci for each year of operation of a 1000-MW plant. This compares to 375,000 Ci ^{85}Kr from the same reactor over the same period, but the biological implications of the tritium are far more complex. Tritium, an isotope of hydrogen, readily exchanges with the body's normal hydrogen. It can, however, escape again by a similar exchange. We thus have a situation where a person exposed to tritium for a short time will absorb some of it via skin, lung, or food and then gradually lose it. The rate of loss, in the absence of fresh exposure, involves elimination of half the existing tritium every 12 days. The concern, therefore, is for continued release of high levels of tritium sufficient to change the natural balance in a significant way. Reprocessing will release all of the tritium contained in spent fuel, some in the form of tritiated water and some in gaseous form, the percentage of each varying with the particular process system selected. Reprocessing all the fuel from the world's operating reactors in the year 2000 (assuming the operation of 700 reactors at an average of 1000 MW each) would have the potential to add about 14 million curies per year (14 MCi/yr) to the world's inventory of tritium. The natural inventory of tritium from cosmic ray interactions and other natural sources is believed to be about 26 MCi, so this would be an undesirable impact. It is clear, therefore, that if worldwide reprocessing is adopted, steps will have to be taken to contain and control much of the associated tritium.

In the United States almost all nuclear fuel processing to date has been in government installations producing plutonium for nuclear weapons. The fuel irradiated in the nuclear reactors at Hanford, Washington, and Savannah River, Georgia, has been reprocessed for its contained plutonium, and the wastes have been collected as concentrated liquid solutions that have been stored in large underground tanks. On the other hand, since 1963 at the Idaho Chemical Reprocessing Plant in Arco, Idaho, the reprocessed waste from the fuel from the navy's nuclear fleet has been converted into dry salts and oxides by heating to very high temperatures (calcining). The French reprocessing plant at Marcoule has gone a step further and since 1978 has converted its high-level radioactive waste to a complex solid glass. The conversion of high-level radioactive wastes to solid form is thus a well-demonstrated technology, although further improvements are still possible and laboratory work continues. Nuclear opponents frequently urge the slowing down of nuclear power development until the United States has demonstrated successful large-scale solidification of nuclear wastes and their final disposal. The technology is clearly available, however, and the problem is getting it installed rather than getting it developed.

The high-level wastes consist of many billions of curies of a wide variety of different fission products and, as we have discussed above, may be in the form of concentrated liquid solutions or of solidified glass or other similar solid form. Since most spent fuel has been stored for at least 150 days before reprocessing, the short half-life isotopes (less than about 15 days) will have decayed away. The remaining material will have half-lives longer than 15 days and up to thousands of years. Remember again, however, that long half-life means less intense radioactivity.

Our ability to contain these wastes for short times in liquid form has been demonstrated at Hanford and Savannah River, although leaks have occurred in some of the older tanks. In such cases the radiation can slowly diffuse down through the ground and possibly eventually reach the water table. At Hanford the rate at which permeation through the soil has occurred has been measured, and it is very slow. There is also safety in that the further the radioactivity moves, the more dilute it becomes. Any leakage such as this is undesirable, however, and as a consequence waste management policy calls for future solidification of all high-level waste before final disposal. The solidified highly radioactive material generates considerable heat because of the energy released in its radioactive decay. The amount of heat generated decreases with time, however, so as long as the initial storage conditions can accomodate the heat generation, there should be no problems later on.

The expectation is that final disposal will be through deep burial in stable geologic formations not expected to change for thousands of years. Most likely areas are deep permanent salt strata such as are found in Kansas and elsewhere, granite rock formations, or possibly basalt formations resulting from ancient lava flows. There seems little doubt that formations exist with adequate stability to protect all future generations from any consequences of the buried radiation. The amount of radiation from any buried batch decreases constantly with time. Although some of the radiation will persist for thousands of years, the isotopes existing longest have the least intensity. Also the physical quantity of material to be buried is relatively small. It has been estimated that the solidified high-level waste from all the U.S. nuclear power production to the year 2000 would fit in a cube 50 feet on each side (125,000 cubic feet).

Our task is to proceed with selection of suitable sites, choose and implement the best solidification technique, and begin the final disposal process. The plan for accomplishing the task was spelled out in the Nuclear Waste Policy Act passed in 1982. Under this act the Department of Energy was directed to prepare a plan to ensure that a safe geological repository will be operational by January 1998. A draft plan was issued in May 1984 aimed at meeting this goal. In December 1984 nine specific sites were identified and draft environmental assessments were released. Three of these sites (Hanford, Washington; Deaf Smith County, Texas; and Yucca Mountain, Nevada) have been picked for further detailed study, and of these one will be recommended to the president as a final choice, scheduled for 1991. The operation goal is still 1998. Steps are also under way to select a second high-level waste repository site. Twelve potential sites in seven states (Georgia, Maine, Minnesota, New Hampshire, North Carolina, Virginia, and Wisconsin) were named in January 1986 for initial consideration. This field is to be narrowed to three by 1991, and if construction is finally authorized the second site should be available by 2006. It is certain that there will be many local objections to the preferred sites and much legal maneuvering to prevent their final selection. It is the clear intent of the government, however, to see that suitable sites are finally designated and that safe disposal of wastes becomes possible in them.

The problem in final selection appears to be more political than technical. There is a natural tendency for residents near a proposed burial site to be fearful of its consequences. Opponents of nuclear power work on these fears to develop local opposition to the disposal project. Site selection has therefore become increasingly difficult, and it will take intelligent and careful education at local, state, and federal levels before residents will see the reality that a burial site will not have harmful consequences and is essential in the national interest.

Commercial nuclear fuel is reprocessed in England and France, but for largely political reasons reprocessing of commercial nuclear fuel is not now carried out in the United States. The spent-fuel elements are currently being stored underwater in large storage pools at the reactor sites or at the one commercial spent-fuel storage site at Morris, Illinois. The Department of Energy is now proposing Oak Ridge, Tennessee, as the site for a Monitored Retrievable Storage (MRS) site. Here, with costs borne by the utilities, a maximum of 15,000 metric tons of spent fuel elements would be stored in a facility with ready access for monitoring radioactivity levels and for removal of fuel elements for reprocessing or final storage at a later date. This would permit decay of much of the short-lived activity before final storage and would keep open the option of reprocessing to recover useful fuel values, permit operation of breeder reactors, and greatly reduce the volume of material for final storage.

The safe disposal of nuclear wastes can be accomplished either as spent-fuel elements or as concentrated solidified high-level waste. The issue of whether to reprocess or not must be settled largely on the basis of balancing the risk of weapons proliferation because of the increasing availability of plutonium against the benefit of an enormous increase in our energy supply. To address that issue adequately would take a book in itself, but the posing of the issue brings us directly to the next major topic in our look at radiation, namely the impact of nuclear weapons on our radiation environment.

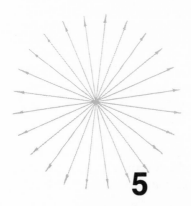

5

NUCLEAR WEAPONS

THE FIRST NUCLEAR WEAPONS TEST at Alamogordo, New Mexico, on 16 July 1945 ushered in a complete new chapter in the story of man's inhumanity to man. It opened the way to a heretofore unachievable level of total destructiveness making insignificant even the bombings of Coventry or the fire storms that destroyed Hamburg and Dresden in Germany. Even the limited and controlled tests that have occurred to date have changed significantly our total radiation environment, and any massive use in anger would leave a heavy imprint of radiation on the world forever.

HOW THEY WORK AND WHAT THEY DO

To understand what *has* happened and what *can* happen we must first learn the elements of what a nuclear weapon is, how it works, and what its effects are. As with nuclear reactors we can start with the phenomenon of nuclear fission in which the splitting of the nucleus of ^{235}U, ^{233}U, or ^{239}Pu simultaneously releases a large amount of energy and several additional neutrons to initiate more fissions. In a nuclear reactor our objective is to control the rate of release of energy so as to provide the benefits of a constant flow of heat for a long time. In a nuclear weapon the objective is to achieve the maximum possible energy release in the shortest possible time.

As children we became aware that powder from a firecracker if ignited in the open merely flashed up in a brilliant flame but without an explosion. To get a big bang the powder had to be contained in the tight jacket of the firecracker itself or perhaps better still inside a tin can as well. Large explosions are the same. They have to be contained or restrained as long as possible to generate maximum pressure before finally bursting out. Thus not only is very rapid generation of energy required, but you need to compress and contain it as it builds to its maximum strength. The larger the source of energy, the greater is the containment force required. This is one of the essential features of a nuclear bomb in common with all other types of bombs.

A unique feature of nuclear explosions involves the concept of criticality. A very small amount of pure plutonium-239 or uranium-235 will lose so many neutrons to the space beyond it that the nuclear chain reaction cannot be sustained. In other words, from each fission event there will be on the average less than 1 additional neutron available for another fission, so the amount of energy released cannot increase. There can be no chain reaction. As we consider larger and larger pieces of plutonium, however, there will eventually be a size at which, since neutrons are only lost from the surface, the rate of neutron loss will just equal the rate of generation. This is known as the "critical mass" or "critical size." Above this size, neutron multiplication increases extremely rapidly. Below it there is none.

A piece of plutonium below the critical mass can be safely held in one's hand, but if 2 such pieces are suddenly put together and add up to a size *above* the critical mass there will be an instant and enormous generation of energy. This is the basic principle of a nuclear bomb. It is the sudden assembly of a larger than critical mass from previously subcritical masses. There must also be a way to contain the energy buildup long enough to generate the incredible explosive power of a large nuclear weapon.

The specifics of how this is done are still secret, but two fundamentally different principles have been used. In one type, 2 or more subcritical masses are fired at each other to form the supercritical mass, and large surrounding masses of heavy material such as natural uranium are used to provide the containment pressure. These are often referred to as "gun type" weapons, a term *not* meaning they are used as artillery shells. The second type involves another new term, "implosion." In this type a subcritical piece or carefully spaced pieces of fissionable material are surrounded by explosives focused inwards towards the fissionable material at the center. When the explosive charge is fired, its force is not dissipated outwards as in an explosion but is concentrated inwards, i.e., "implodes" on the fissionable mass at the center. This is compressed from all sides to become a supercritical mass and instantly generates a huge quantity of energy. The initial implosion acts to contain the nuclear explosion long enough to let it achieve its enormous maximum force.

A nuclear weapon, whether gun type or implosion, unleashes three different destructive forces: blast, heat, and radiation. Each has different characteristics and for each there is a different defense. In a typical nuclear explosion the immediate damage is largely the result of blast. The shock wave can destroy buildings, bridges, utility poles, and wiring or any other structures in its path for a distance of 1 to 2 miles from the explosion center for a nominal 20-kiloton weapon such as used against the Japanese at Hiroshima or Nagasaki. (The size of a nuclear explosion is defined by indicating the amount of TNT required in a chemical explosion to generate a blast of equal force.) The much more powerful weapons available today can do proportionately greater damage.

Thermal radiation (heat) is also a major factor through its capacity to burn exposed skin and to ignite fires that continue their destruction long after the blast has subsided. The thermal radiation results from the extremely high temperatures achieved in the fireball generated by the huge release of energy. Temperatures achieved are in the order of 300,000°C, and the radiant energy from this miniature sun lies in the spectrum from ultraviolet light through visible light and into the infrared, which is converted to heat when it is absorbed. The pulse of thermal energy actually follows the blast, but immediately behind it. The thermal radiation pulse lasts for about 3 seconds and generates in that brief time enough heat to char wood, ignite fires, and cause body burns at distances up to a mile or more from the detonation point of a 20-kiloton weapon.

The nuclear explosion also generates large quantities of gamma (γ) radiation, a by-product of the fission events. These radiations are highly penetrating; even heavy clothes, protective against thermal radiation, do nothing to screen out γ radiation. We have pointed out that a radiation dose of 400 roentgens will be fatal to half of those receiving it. The other half will be very ill but will recover. This level of dose from γ radiation occurs about 4200 feet from the explosion site of a 20-kiloton weapon. Closer in, much higher and obviously lethal doses occur. Farther out, the dosage is less. Gamma radiation is thus a significant factor in casualties in the range from 2000 to 9000 feet from the explosion of such a weapon. At shorter distances blast and thermal effects are so immense that there is no survival anyway, and beyond 9000 feet the γ radiation has dropped below dangerous levels. Within this critical range between 2000 and 9000 feet, however, shielding can be very important. About 20 inches of concrete, for example, could reduce a dose of 400 R to 25 R, a dose that might produce illness but probably would not kill. A trench, a wall, a large tree if directly in line with the explosion, could be a lifesaver in some cases.

The final instant hazard in a nuclear explosion is the huge burst of neutrons also emitted just as the explosion occurs. A neutron pulse containing 10^{11} neutrons per square centimeter is generally considered lethal to humans. With a 20-kiloton explosion, such as we have been considering, this level of neutrons is still present about 2000 to 2400 feet from the center of the explosion. Beyond this distance the neutron intensity falls off rapidly. We have already noted that γ radiation is 50% fatal out to 4200 feet. The effect of neutrons in a Hiroshima- or Nagasaki-type weapon is thus swamped out by the more dominant effects of blast, heat, and γ rays. There has been much talk in recent years of the existence of a "neutron bomb" that maximizes the lethal effect of neutrons and minimizes the effects of blast. How this has been accomplished has not been made clear, nor is it clear whether γ rays are also instrumental in its effects.

In the discussion so far we have not considered the hydrogen bomb or fusion bomb as it can also be called. In this weapon the nuclei of very light atoms, hydrogen and its isotopes deuterium (^{2}H) and tritium (^{3}H), are fused together and in the joining release huge amounts of energy, even more than in the splitting of uranium or plutonium nuclei. Blast and thermal effects from a fusion weapon will be similar in nature to those from a fission bomb, but a pure fusion weapon would not have a high-neutron yield nor would it leave behind the wide variety of highly radioactive products typical of a fission weapon. It seems probable, however, that most fusion weapons have a fission trigger or get some of their energy from fission reactions and are in fact hybrid weapons. Their effects, while different in degree, may not differ greatly in nature, and our discussion from here on will focus on the effects of fission weapons only.

The instantaneous effects of nuclear weapons are largely, as we have said, from blast, from heat, and from γ radiation and neutrons. The damage they can inflict is enormous and horrible. At Hiroshima 100,000 people died from a single bomb and at Nagasaki 40,000 more were killed. These, mind you, were the casualties from 20-kiloton weapons, and today 100-megaton weapons are available, 5000 times as powerful. Compared to a 20-kiloton bomb such a weapon would have similar blast effects 17 times as far away, so it would destroy buildings up to 17 to 34 miles from its point of explosion. The effects of heat from the fireball become increasingly significant in larger and larger weapons; for a 100-megaton monster, third-degree burns and the starting of fires would occur even up to 70 miles from the explosion site. The lethal range for γ rays, however, would be increased from a little less than a mile to only about 2 miles. Thus the larger the weapon, the more the blast and thermal effects will be the predominant destructive force at the instant of the explosion.

RADIATION AFTEREFFECTS

The damaging effects of *radiation* associated with a nuclear weapons explosion will largely come after the incredible destruction at the instant of the blast. The pattern of the effect will vary depending particularly on the height of the burst. An explosion several thousand feet above the ground will maximize the effects of blast and fire. The enormous heat generated on the ground will spawn huge winds that will rush in towards the hottest spot directly under the explosion and then rise up in a great updraft carrying with it the radioactive fission products liberated at the moment of the explosion. The towering mushroom cloud so typical of a nuclear explosion will climb up into the stratosphere 10 miles or more above the earth. There the upper atmosphere winds will disperse the fine radioactive particles and spread them above the earth following the vagaries of wind and weather. These fine particles may stay aloft for days, weeks, months, or even years, but over time they will settle back to earth as highly diluted radioactive fallout. It is not often recognized that at Hiroshima and Nagasaki there was hardly any fallout of radioactive particles on the cities themselves. Essentially all the fission products followed the path described above and became dispersed in the upper atmosphere.

An explosion on or near the ground will have quite a different pattern. In this case enormous amounts of dirt and debris will be sucked up into the initial fireball, and most of the fission product radioactivity will be intermingled with it and adsorbed on it. Much of this debris will be too massive to rise into the stratosphere, but instead will be blown downwind from the explosion site and will settle out or be carried down by rain in paths perhaps hundreds of miles long. Minor amounts of radioactivity produced by absorption of neutrons in dirt and structural debris will be added to the radioactive burden in the dust cloud. The total amount of radiation in such a cloud will depend on the size and number of bombs exploded, the height of their explosion, the nature of the target, and the weather conditions at the time.

With all of these variables it is not possible to make realistic estimates of the casualties that might result from radioactive fallout from a ground level burst of a large weapon. One attempt, however, concluded that a ground level burst of a 20-megaton (20,000 kiloton) weapon could contaminate an area of 4000 square miles with enough fallout, if evenly distributed, to kill half the people in the area if they remained there unprotected for 48 hours. Let's think about that in realistic terms for a moment. We are probably thinking about an area perhaps 40 miles wide and 100 miles long downwind from the explosion. Over the first 10 to 15 miles out from the explosion center there would be massive blast and fire damage in addition to the radiation, but at greater distances the radiation would be the major hazard. We can assume that within a fairly short time, a few hours perhaps, the people not seriously injured would learn what had happened and recognize that a severe radiation hazard existed. There would be two alternate strategies they could use to try to save themselves. The first option would be to get out of the area just as fast as possible, preferably by car with all windows shut tight. The objective would be to take the shortest and fastest route to an area outside the fallout cloud. This would require knowledge of the extent and direction of the cloud, but one would hope there would be radio reports to learn the best routes. Once outside the fallout area a thorough bath, clean clothes, and a hospital check would be in order. Flight such as this would pose severe problems to the uncontaminated areas, however, since each fugitive would carry with him on car or clothes some of the very radioactivity he was trying to escape.

The second option for survival would be to remain indoors with windows and doors tightly closed and preferably in a basement where concrete walls and the building above would provide good shielding. The intensity of radiation outside would decrease rapidly after the fallout cloud had passed. Short half-life intense radioactivity would disappear quickly and only the radiation from longer half-life isotopes would remain, decaying slowly away. The rate of decay can be shown by considering a point where an unprotected individual in the second hour after the explosion would receive 2500 roentgens (R) (enough to cause early death). One day after the explosion an individual standing at the same point for a whole day would receive 950 R or an average of 40 R/hr. A month later someone at the same point for a whole month would receive 220 R or a little over 7 R/day or 0.3 R/hr. Thus those remaining in a protected location would know that the radiation levels were constantly decreasing outside. Their critical question would be when to leave the shelter area and try to escape to a completely uncontaminated region. This decision would be made clearer if the radiation levels in the shelter were known, but not many of us own Geiger counters. The choice would have to be made on the basis of availability of uncontaminated food and water, the physical health of the people involved, and knowledge of the radiation levels outside. Other considerations would be the availability of adequate transportation and the extent to which the transportation was contaminated with radiation. Was the car in the garage or outside, for example. Common sense would have to dictate, but the best time would probably be 1 to 3 weeks after the explosion. Probably by then medical check-in centers could have been established for decontamination and health checks and the beginning of adequate medical care.

By either strategy and the intelligent use of knowledge of the characteristics and effects of radiation many lives could be saved and damaging health effects minimized. It is probable, therefore, that the total casualties in death and sickness from fallout, although major, would be less than those from immediate blast and fire. Radioactivity and fallout may be only the frosting on the devil's cake.

LONG-TERM EFFECTS

We cannot leave the subject of nuclear weapons, however, without considering their long-term effects. Like nuclear reactors they add to the world's background of radiation, and we need to understand the nature and magnitude of that addition. It can be approached through consideration of measurements made of the radioactivity resulting from the many nuclear weapons tests carried out in the atmosphere particularly in the period between 1950 and 1963.

WEAPONS TESTS: 1950–1963. The era of nuclear weapons began on 16 July 1945 with the first nuclear test explosion, the "Trinity shot," at Alamogordo, New Mexico. The wartime use of the weapons at Hiroshima and Nagasaki was followed by a period of limited testing at the Pacific Ocean atolls of Bikini and Eniwetok. Five weapons were tested there prior to 1950. Then in December 1950 President Harry Truman approved the use of a test site in a sparsely settled area of Nevada. His goals were ease of access and minimization of cost, and his advisors felt the risks to inhabitants of the area were minimal. In the section on Weapons Test Fallout (p. 122) we will see whether or not that advice was sound. In any event, between 1951 and 1958, 121 nuclear weapons were exploded in the air at varying heights above the Nevada desert.

By 1958 there was growing concern about the effects of radioactivity dispersed into the atmosphere by these tests. The Russians had conducted their first atmospheric test in 1949, and they had also continued an active test program thereafter. In 1958 Russia and the United States agreed to a voluntary ban on further tests of nuclear weapons. This agreement expired in 1961, and both sides resumed testing at an accelerated pace until the Limited Nuclear Test Ban Treaty was signed on 5 August 1963. In that brief period of a little less than two years the United States alone exploded 102 nuclear weapons in the atmosphere, including both fission and fusion weapons. After 1963 all weapons testing was carried out underground by both the USSR and ourselves. For those concerned about deep burial of nuclear wastes, it is sobering to recognize that between 1963 and 1980 the United States conducted 316 underground nuclear tests. The radioactivity from those tests buried under the

Nevada desert is already equivalent to the wastes from many nuclear power reactors.

Not all countries have adhered to the practice of keeping all nuclear weapons tests underground. Both France and China have continued the practice of atmospheric tests and thus have continued to add to the airborne burden of radioactivity. The number of tests in the atmosphere after 1963 has been far less than the combined programs of the USSR and the United States before that year, however. Measurements of radio-activity in the atmosphere show this fact dramatically, with continually increasing levels until 1963 and gradually decreasing levels thereafter as a result of radioactive decay.

With an understanding of this pattern of testing we can now turn to the measurements themselves and what they tell us about the radioactive imprint of nuclear weapons on a global basis. It isn't practical to consider all of the radioactive isotopes injected into the air by a nuclear explosion. Some have half-lives so short that they vanish within minutes or hours. Some are created in quantities too small to be significant. There are a few, because of their long half-lives and the way they react in the human body, which are the most important and can be followed as examples most instructively. The ones we have selected are tritium (^{3}H), carbon-14 (^{14}C), strontium-90 (^{90}Sr), cesium-137 (^{137}Cs), and plutonium. Krypton-85 could also be considered, but its major source is from fuel reprocessing rather than weapons tests (see Chapter 4).

TRITIUM. We discussed briefly the characteristics of tritium in Chapters 3 and 4, but it won't hurt to remind you that this isotope of hydrogen has a 12.3-year half-life and emits an 18 keV β particle. You should recognize by now that this is a weak β and not highly damaging. Small amounts of tritium are generated in a fission explosion, but very large amounts are generated in the explosion of a fusion weapon. This is probably the largest source of tritium in our environment today, even though there is a continuous natural production from cosmic radiation. Most of the tritium produced ends up in water. The familiar chemical expression for water containing two atoms of hydrogen and one of oxygen, H_2O, becomes HTO or a molecule with one atom of hydrogen, one atom of tritium, and one atom of oxygen.

The amount of tritium in a given sample of water can be measured very accurately, and starting in the early 1950s such measurements have been made on a continuing basis over many parts of the globe in lakes, in rivers, in the oceans, and in the air. The pattern that they show is fascinating. In the United States, for example, the base concentration of tritium in natural water was perhaps 15 to 20 pCi/l (picouries per liter). After the start of weapons testing this gradually increased until peaks of a few thousand picocuries were reached in 1963. By 1974 this had decreased to slightly over 100 pCi, still almost 10 times natural background, and was continuing on downwards.

To understand the global spread of a volatile isotope such as tritium we have to learn a little of the jargon of the atmospheric scientist. Two important terms are stratosphere and troposphere. The stratosphere is the upper atmosphere from 7 to 15 miles up. Once a gas or very fine particle reaches this level it tends to stay there and drop into the lower atmosphere region, the troposphere, only very slowly. It can move about horizontally, but material injected in the northern half of the globe doesn't readily diffuse past the line of the equator into the southern half. Levels of tritium measured in the Southern Hemisphere of our globe have been only one-tenth or less the levels measured in the Northern Hemisphere, because almost all the weapons tests have been made north of the equator. There are seasonal effects also, and it has been shown that transfer down from the stratosphere to the troposphere occurs more rapidly in late winter and gives rise to a maximum fallout rate down to earth in the early spring. Transfer from the troposphere to ground is more rapid because the troposphere is the atmospheric region below a height of about 7 miles, and in this region clouds form and rain falls carrying down with it any fine particles floating in that region of the air.

The tritium in the world originating from weapons explosions has followed one of two paths. Much of it was first injected into the stratosphere at the time of the explosion and may have remained there for even a few years before making its way down to the troposphere and then being carried down to earth in rain. The other portion remained in the troposphere after the explosion and, following the local wind and weather pattern, came back to earth within hours, days, or weeks in areas perhaps up to a few hundred miles from the explosion site. The amount that followed each path would depend on where the bomb was exploded, how big it was, and how high above the ground it was detonated.

The wisdom of banning atmospheric tests of nuclear weapons has been amply demonstrated by the measured pattern of tritium distribution in the world and the gradual reduction in its level since the 1963 cessation of atmospheric tests by the United States and the USSR. A similar move by France, China, and others would be most welcome. It is not that even the measured amount of tritium has caused visible harm in sickness or death. A picocurie, you will recall, is one trillionth of a curie, so even a thousand picocuries is only one billionth of a curie of radiation. We do not really know whether such small amounts of tritium are biologically harmful. If they are, the harm is very small since we have not been able to identify it among all the other hazards of living.

CARBON-14. Carbon-14 (^{14}C) is another important isotope generated by nuclear weapons. It emits only β radiation at a fairly intense 0.16 MeV and has a half-life of 5730 years, so once formed it remains with us for a very long time. In a nuclear explosion it is formed by the interaction of neutrons with the nitrogen in the air, and estimates are made that man-made ^{14}C (mostly from weapons explosions) introduced into our atmosphere up to 1972 amounted to about 6 million Ci. ^{14}C formed naturally by cosmic rays is about 38,000 Ci/yr. Thus the weapons testing between 1945 and 1972, a period of 27 years, produced as much ^{14}C as would be produced naturally in 158 years. This is probably not a very significant pulse in the level of ^{14}C in the world. Measurement in tree rings and in ocean and lake sediments have shown that over a period of 10,000 years fluctuations in ^{14}C level of up to 10% have occurred naturally. Man's perturbation is thus larger than nature's, but not by very much.

Even these small perturbations must be taken seriously, however, because ^{14}C is incorporated into the body in proportion to its existence in the atmosphere. Plants absorb the carbon dioxide (CO_2) in the air, and animals and man eat the plants. With a time lag of about a year, the level of ^{14}C in man will match the level in the air. This radioactive carbon is naturally present even in the DNA molecules of our genes, and changes in the DNA can create changes in the pattern of our hereditary characteristics. Will an increase in the level of ^{14}C affect the rate or nature of genetic change? Again we do not know, because the differences are tiny and we cannot clearly separate out any effect. It would seem wise to minimize this tampering with nature's balance unless we can see a real and positive gain thereby. In human terms the proliferation of nuclear weapons hardly seems such a positive gain.

STRONTIUM-90. Strontium-90 (^{90}Sr) is important because of its half-life and emission characteristics and because it easily gets into the human food chain. It emits only β particles with energy of 0.54 MeV with a half-life of 28 years. It decays, however, to yttrium-90 (^{90}Y), which has a half-life of 15 hours and emits a high-energy β particle of 2.27 MeV. Strontium in soil is absorbed by growing plants that may then be eaten by animals or man. Strontium in the ocean is picked up by marine plant life and from there a small percent of it reaches fish that may be food for man.

About 2½% of the fission products produced in a nuclear explosion are ^{90}Sr, so it is one of the commoner isotopes produced and has considerable potential for harm. Its path in the atmosphere is the same as the other isotopes, and it has been distributed worldwide through repeated injections into the stratosphere in weapons tests. The amount depositing on the earth peaked in 1963 at nearly 3 million Ci, but it was deposited very broadly in enormously diluted although still measurable form. Since it has a half-life of 28 years and it is now (1986) 23 years since cessation of most weapons testing in the atmosphere, the total amount of ^{90}Sr radioactivity spread broadly in the world is decreasing. The cumulative amount in existence on the ground or in oceans, lakes, and rivers peaked at about 12 million Ci in 1966 and has been decreasing ever since. About 80% of it is in the Northern Hemisphere since most of the weapons tests were in that portion of the globe.

There has been intensive study on a worldwide basis for many years to determine exactly where ^{90}Sr has accumulated in the world and what its pathway is to man. Strontium behaves in body chemistry much like calcium, which is a major part of bone and teeth. Milk is one of the best sources of calcium for man, and milk is also one of the commonest paths by which man absorbs ^{90}Sr. The field plants absorb strontium from the soil through the roots. The cows eat the grass, and the strontium goes with calcium into the cow's milk and hence to us. Measurements of ^{90}Sr content in cow's milk have therefore been made regularly in many places to monitor it for our health and to aid in our understanding of how ^{90}Sr is spread in the world. Since the amounts are so tiny it is impossible to measure the amounts of strontium chemically. Even single particle radiation is theoretically detectable, however, so the radiation level in calcium separated from the milk can be used to assess the amount of ^{90}Sr. It is usually reported in picocuries per gram (pCi/g) calcium, and we will use those units here. In 1963 measurements at many points in Europe and the United States ranged from 16 to 28 pCi/g, with a middle value of 23 pCi. As we have said, a picocurie is not one thousandth or one millionth or one billionth of a curie but one trillionth of a curie, a tiny, tiny amount only measurable because of our unique capabilities for detecting radiation. Are these amounts significant? Probably not, but we will grapple with that issue in Chapter 8.

By 1975 the amounts measured in milk in the same areas had dropped to 3 to 5 pCi and are presumably less today. Graphs have been made showing how the amount of ^{90}Sr has varied with time in vegetables, fruit, grain, and meat. All show the same pattern of major increases up until 1963 to 1964, followed by initially rapid and then slower decline. These measurements clearly establish the relationship between explosions of nuclear weapons in the atmosphere and the worldwide spread of radioactivity from them.

CESIUM-137. The pattern for cesium-137 (^{137}Cs) is very similar to that for ^{90}Sr. It has a similar half-life (30 years), but has the added complication of emitting 0.66 MeV γ rays as well as β particles with maximum energies of 1.17 MeV. Cesium follows the body chemistry of potassium (K), so 80% of it ends up in muscle rather than in bone. It, too, is absorbed in plant life and moves from there to man. The highest levels incorporated with normal potassium are encountered where meat is a predominant portion of the diet such as in the sub-Arctic regions in Canada, Alaska, and the USSR. Levels as high as 10,000 to 20,000 picocuries per gram of potassium (pCi/gK) have been observed in 1963 to 1964 in these regions. This reflects the ready absorption of ^{137}Cs in the lichens that are the predominant food for reindeer or caribou, in turn a staple diet for northern man. More typical levels in the middle latitudes, areas such as France, Germany, Japan, or the United States, range from 100 to 300 pCi in the maximum years and were down to 5 to 20 pCi by 1975.

PLUTONIUM. Finally we come to plutonium, often categorized in the press as "one of the most toxic substances known to man." This is a considerable exaggeration; but plutonium is a dangerous material in sufficient quantity, as are other radioactive isotopes we have discussed. The plutonium isotope of most concern is ^{239}Pu, which has a half-life of 24,000 years and emits high-energy α particles (about 5 MeV). Because of the long half-life the rate of emission is relatively low, but in the body plutonium moves to the bone where its radiation can affect the formation of red blood cells. Relatively small amounts can therefore be harmful in the long term. About 60% of the plutonium dispersed in a nuclear weapons explosion is ^{239}Pu, but there are also significant quantities of ^{238}Pu, ^{240}Pu, and ^{241}Pu with half-lives of 87 years, 6600 years, and 14 years, respectively. All but ^{241}Pu are α emitters, but the latter is a β emitter with some associated γ activity and decays to americium-241, an α emitter that has a 430-year half-life and also emits an appreciable amount of 60 keV γ radiation. The amount of ^{241}Pu in any batch of plutonium varies depending on the origin and history of the particular batch. It is formed by the absorption of a neutron in ^{239}Pu to form ^{240}Pu, which in turn, on absorbing another neutron, becomes ^{241}Pu. The constant generation of neutrons in a nuclear reactor permits a gradual increase in ^{241}Pu as the fuel remains in the reactor. The longer a fuel has been irradiated, the higher is the percentage of ^{241}Pu. Thus plutonium separated from long-term reactor fuel irradiation will have a higher concentration of ^{241}Pu than that separated after a short irradiation. The plutonium used in weapons is usually separated out after relatively short periods of reactor irradiation and hence is fairly low in ^{241}Pu content.

Plutonium that has just been chemically purified in nuclear fuel reprocessing will have very little γ activity and will emit mostly high-energy α particles. In the time after its purification, however, the ^{241}Pu will be decaying with a half-life of 14 years and building up the decay product americium-241 (^{241}Am) which, as we have indicated above, has a half-life of 430 years and gives off both α and γ radiation. Because of this the amount of radiation from a freshly purified sample of plutonium will gradually increase over many years. Let's suppose, for example, that the initial sample of plutonium contained 6% ^{241}Pu. In the first 14 years, one-half of this would be converted to ^{241}Am, almost all of which would still be present in view of its long half-life. So at the end of 14 years our initial sample would contain 3% ^{241}Am, but only 3% ^{241}Pu because half of the original 6% would have decayed away. After 28 years there would be 1.5% ^{241}Pu and 4.5% ^{241}Am, after 42 years (3 half-lives of ^{241}Pu) there would be 0.75% ^{241}Pu and 5.25% ^{241}Am, and so forth until after 100 years or so almost all the ^{241}Pu would be converted and no further increase in radiation levels would occur. After that time the radiation levels would decrease as the ^{241}Am decayed away to neptunium-237.

Plutonium from weapons tests follows a fallout pattern similar to the ^{90}Sr we discussed earlier. It is estimated that about 320,000 curies of ^{239}Pu had been injected into the atmosphere by 1973, and there has been a small additional amount since. This is the equivalent of several *tons* of plutonium-239 dispersed about the globe. There has also been 26,000 Ci ^{238}Pu released into the atmosphere, but 17,000 Ci came from a unique source, a vagrant satellite. ^{238}Pu can be used as a pure source of high-energy α emission to tickle semiconductors into generating small amounts of electric power. Such sources are used to provide power to satellites, and in 1964 one of these satellites reentered the earth's atmosphere and was burned up, thereby releasing its ^{238}Pu as finely dispersed plutonium oxide.

Studies conducted in New York showed that in the period from 1972 to 1974 an average New Yorker ate food containing a total of 1.6 picocuries of radiation in the form of plutonium-239 and 240. Similarly it has been estimated that in the period between 1954 and 1975 the average New Yorker breathed in 43 pCi plutonium. This is the legacy of weapons testing in the atmosphere. Medical research has shown that 40,000 pCi plutonium in the body will not produce any evident harm, so we are still safe by a factor of 1000. Nevertheless plutonium is a new source of radiation in our environment that requires realistic control.

SUMMARY

It is clear from all the discussions above that nuclear weapons in war and in test have been responsible for major changes in the radiation environment of the world. The immediate effects of only 2 small weapons have caused almost total destruction of 2 medium-sized cities with 140,000 deaths. The long-term effects of weapons tests in the atmosphere have been measurable, although not evidently harmful, all over the world. There have been perhaps 400 atmospheric weapons tests in all. These have been responsible for most of the world's uncontrolled inventory of tritium, strontium-90, cesium-137, plutonium, and a variety of radioactive isotopes of lesser importance. They have made a significant addition to the world's inventory of carbon-14. While the true horrors of nuclear war would be in the immediate effects of blast and flame, the local radiation residue from ground-level explosions would be disastrous as well. The long-term effects from upper atmosphere dispersal of radioactivity would only be clearly harmful if many thousands of weapons were exploded.

In that case other side effects would also become major factors. Recent studies (1983) have shown, for example, that the huge clouds of smoke and soot that would rise from the explosions of thousands of megatons of nuclear weapons could result in absorption of solar energy in the stratosphere and upper troposphere causing increases in temperature at these altitudes of up to 90°C (162°F). On the earth below, however, where the sunlight couldn't penetrate, average temperatures would drop by up to 30°C (54°F) causing havoc to crops and vegetation. Thus a nuclear winter could be imposed on an already devastated earth, crippling the production of the agricultural food supplies on which any revival would have to depend.

The survival of civilization as we know it today depends on our preventing the use of nuclear weapons on any major scale. The total chaos that would be created goes far beyond questions of radiation alone and encompasses as well the effects of blast, heat, and major environmental changes that might be induced.

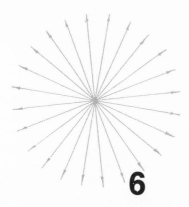

6

NUCLEAR MEDICINE

WE TURN NOW FROM THE HARMFUL SIDE of radiation to its potential for good in the diagnosis and treatment of illness. Almost all kinds of radiation are used in some fashion: X rays for diagnosis and cancer treatment, laser light for eye surgery, microwaves for diathermy, radioisotopes to track body chemistry or destroy unwanted cells, γ rays for cancer destruction, electron beams to treat skin lesions, and many more special applications. In this chapter we will enlarge on the most common uses, but many more exist which we will not touch upon.

DIAGNOSTIC X RAYS

Nearly everyone who reads this book will have had some form of diagnostic X rays taken during his or her lifetime. Examples are dental X rays, chest X rays, the familiar and much disliked gastrointestinal or GI series, X rays of a broken bone, breast X rays (mammography) for detection of cancer, etc. In these tests a beam of X rays is passed through the part of the body being diagnosed and registers on a film placed behind it. "Bite down" says the technician in the dentist's office as the tab on the film holder is slipped between your teeth. A short "buzz" of the machine, out comes one film holder and in goes another. "Buzz, buzz" and the job is done. In the process you may have absorbed several hundred millirads (mrad) of radiation on your cheek for each exposure. "Take a deep breath" says the technician taking a chest X ray. You stand against the film holder and the familiar voice from the protected booth by the X-ray machine says "Ho-o-o-ld it!" Again the buzz, longer this time, and "You can breathe now," and the picture of the expanded lung is complete. Exposure to your chest has probably been 20 to 30 mrads.

The picture quality is based on the different absorbing power of the various parts of the body within the beam. Bone absorbs X rays much more than tissue does, so when a hand is X-rayed, for example, the film receives more X rays through the fleshy part of the hand than it does directly behind the bones of the fingers. The film is developed or darkened in the fleshy regions but is less affected under the bones, and the outline of the bones shows white on the film. The best contrast is obtained when the proper exposure time and X-ray intensity are selected. The thicker or more dense the body section to be traversed, the greater must be the time or intensity or both. Since dental X rays are seeking contrast between bone (teeth) and small cavities, they have a fairly high dose rate, but over a very small area. A chest X ray is seeking to show the relatively slight differences between a cancer lesion, for example, and normal lung tissue. Both are relatively low density and absorb relatively small amounts of X ray; so the intensity may be less, but the area exposed is fairly large.

More complex procedures may give considerably higher doses. To get a good X-ray picture of the upper intestinal tract the patient is usually asked to swallow a thick heavy liquid containing barium salts that provide added density to enhance the contrast. A single exposure may require sufficient X-ray intensity to give a dose of 900 mrads to 2 rads at the skin surface. Exposure to the organs within will be less, but precise measurements can only be made at the body surface and internal values must be calculated from knowledge of absorption rates, beam scattering, and other factors.

X-ray photography is the taking of still pictures one at a time in the same way we take snapshots with our home camera. Sometimes, however, the doctor wants to watch events happening in real time. For this he resorts to fluoroscopy. In this technique the X rays emerging from the body strike a screen that emits light (fluoresces) in accordance with the energy of the impinging X rays. Thus, while the X-ray beam is on, a real time image of the part being X-rayed is seen on the viewing screen. It is like an X-ray instant movie. For it, the X ray must be on for times up to a minute or more, and exposures of 5 to 10 rads or above are common. These are not trivial amounts of radiation and, although the risks are not large and only involve reactions perhaps years later, fluoroscopy should only be used where the reward of accurate diagnosis is worth the risk. This is often the case where an obstruction or cancerous tumor is involved, but fluoroscopy should not be resorted to lightly.

The risk/reward balance in the use of X rays has changed with time. We are more cognizant today of the risks of long-term effects from moderate levels of X rays, but at the same time advanced technology has made possible improved resolution with lower beam intensities. Even so, certain types of diagnostic X rays have been dropped almost completely because the risks were too high for the knowledge gained. This is particularly true in the use of X rays to determine the position of a baby prior to birth. An obstetrical abdominal X ray usually gives a dose of about 300 mrads, and an X ray to determine a woman's pelvic structure would give a dose of perhaps 600 mrads. In view of an apparent high sensitivity of the human foetus to X rays, particularly in the first few months after conception, X-ray exposure in this period is unwarranted unless severe complications make the risk essential. Fortunately new techniques using sound-wave scanning are being developed that appear to be much safer.

Another area of concern has been the impact of head X rays for neurological studies of the brain. In the old fashioned X-ray methods rather high-intensity beams were necessary to penetrate the skull and still have good resolution on the film. This could result in doses to the cornea of the eye of as much as 20 to 80 rads. Repeated examination at this level could lead to cataracts, which are a fogging of the lens of the eye. In recent times a new technique called computerized axial tomography or CAT scanning has been developed. In this technique the film is replaced by an array of tiny sensitive X-ray detectors recording the X-ray intensity at that spot and feeding a signal representing that amount to a computer. The computer takes all the signals from the myriad sensors and reconstructs them into an X-ray picture. By manipulating the electronics the operator can vary the contrast, and since the signals are stored in the computer memory he can continue to do this long after the scan itself has been made.

For the computer to accomplish the reconstruction it has to have signals taken from very brief X-ray beam pulses from many different angles through the head to the sensor. This is achieved by rapidly rotating the X-ray source and receiver and making many very short-time X-ray pulses during the scan. Because of the high sensitivity of the detectors the total X-ray dose from such a procedure is less than from the old film techniques, and the results provide far more detailed information. The technique has been broadened from head scans to whole-body scans, and instruments now exist that produce what are essentially cross-section X-ray maps through any desired section of the body. Remember, if you will, the magician's trick of putting the pretty girl in a long box and then sawing her in two, only to have her reappear whole and unharmed. The whole-body scanner can make an X-ray picture of the same cross section the saw would have gone through and will hold the picture in its memory to be recalled at will or printed in permanent form. The cost of CAT-scanning machines is very high, but their benefit is enormous through improved information and safety.

One cannot leave the subject of diagnostic X rays without considering briefly the relationship between X rays and genetic damage. Recent years have seen enormous strides in the extent of our knowledge of the complex biochemistry through which hereditary traits are passed on via DNA molecules, the intricate helical spiral of C, O, N, and H atoms. We are also beginning to understand better the way in which excessive radiation can disrupt this complex system and introduce unwanted genetic or hereditary change. This may be one of the ways in which evolution has occurred over eons of time. The practical consideration involving diagnostic X rays, however, is the protection of the genetic material from inadvertent harm. Damage can be done through either the female or male genetic links, the egg or the sperm. For this consideration, therefore, the only X-ray dose of concern is that which irradiates either male or female sex glands, the testes or the ovaries (often lumped under the term gonads). We will discuss in Chapter 8 the mechanics of genetic damage and the levels of radiation that may induce it. Here we will consider its implications in X-ray diagnostics.

It is easy to see that a dental X ray is not going to involve the gonad area unless there is a serious leak from the X-ray machine or some very poor operation. Focusing a chest X-ray beam is a little harder, however, and heavy leaded aprons are sometimes used for added protection. X rays of the pelvic region clearly involve considerable gonadal dose and are therefore more serious in this regard. Medical radiation experts will talk of the GSD of a given radiation treatment meaning its "genetically significant dose," and they will try to keep this as low as possible.

In 1970 in the United States there were 112 million visits to medical facilities for X rays and 68 million visits to dental facilities for X rays. While some of these diagnostic tests may not have been necessary, most of them served a useful purpose. They are one of the largest sources of exposure to radiation experienced by the average individual—far larger than the exposure from nuclear reactors or even from weapons test fallout. There is growing and probably proper pressure to reduce the total level of radiation from this source. But the benefits enormously outweigh the hazards which, like those from all low-levels of radiation, are hard to identify precisely although we know they exist.

RADIOISOTOPES IN DIAGNOSIS

Although X rays are by far the most common form of radiation used in medical diagnosis, there is growing and highly effective use of radioisotopes also, particularly those emitting γ radiation at energy levels not strongly absorbed by tissue and having short half-lives. The procedures involve incorporating the radioisotope directly into the tissue of the patient to be diagnosed and then following its progress with sensitive detectors, usually scintillation counters.

The concept is perhaps best understood through discussion of specific examples. One of the simplest is the use of iodine-131 for diagnostic scanning of the function of the thyroid gland. This small gland situated in the neck with a lobe on either side of the windpipe controls the body metabolism (that is, the rate of utilization of food and oxygen and their conversion to active energy and growth). The element iodine is an important part of this chemical control system and is readily stored in a properly functioning thyroid. In cases where thyroid malfunction is suspected, the patient can be given a drink or a capsule containing 50 to 75 μCi of ^{131}I. A normal thyroid contains about 8 mg (milligrams) iodine. There are 1000 mg in a gram and 28 g in an ounce, so the amount of iodine in the gland is a very small portion of an ounce. About 50 μg (micrograms) are replaced in the thyroid each day, and some of the replacement comes from the radioactive iodine that has been administered. Thus within about 2 hours after the capsule has been swallowed, a tiny amount of the ^{131}I will be present in the patient's thyroid.

^{131}I has a half-life of 8 days and emits both a 364 keV γ and a 600 keV β particle. A sensitive scintillation counter placed near the patient's neck can detect the γ rays and when used in a scanning mode can create a map showing relative amounts of ^{131}I in the various areas of the thyroid. These maps can tell the doctors a great deal about how that particular thyroid is functioning and about the nature of any problems that may exist in it. As a result the proper medication or the proper surgical procedure can be prescribed.

Some people are concerned when they realize that in these tests radioactivity is actually incorporated into the body. They should remember that natural radioactivity already exists there in the form of ^{40}K and ^{14}C in particular. They should also realize that the amounts are very small, that the body's natural processes will eliminate much of the radioactive iodine through the urinary system, and that the short half-life ensures that the radiation will be essentially all gone in 80 days (10 half-lives) anyway. The information from the tests has great value, and the possibility of harm is very, very low.

There are many more complicated tests concerning thyroid function that can be made using ^{131}I. Some of this iodine is used in the thyroid gland to produce the hormones regulating the body's metabolism. Samples taken at appropriate spots can be analyzed quantitatively through measurements of amounts of γ radiation emitted. This information can tell the skilled diagnostician much about how the hormones move and function in the body. Wherever it goes ^{131}I is a signaling messenger telling us where it is and what it is doing.

Many other isotopes can be used for this signaling function, since the complex body chemistry naturally makes use of many different chemical elements. Chromium, for example, is an essential ingredient in blood chemistry, and the isotope ^{51}Cr can be used in a variety of ways to evaluate the performance of a patient's blood system. Does he have enough red blood cells? Do the cells survive as long as they should? ^{51}Cr has a 27.8 day half-life and emits an easily detectable 322 keV γ photon. If the doctor wishes to know the total volume of red cells in a patient's blood, he first draws a carefully measured blood sample and adds to it some sodium chromate tagged with a known amount of ^{51}Cr. When the ^{51}Cr is evenly distributed, the sample is returned to the blood stream and, after a waiting period of hours or days, a new blood sample is taken. If the waiting period has been long enough, the ^{51}Cr will have become evenly distributed in the entire blood stream. The volume of red cells in the small sample taken can be exactly measured, and the amount of ^{51}Cr can be exactly measured also by radiation detection instruments. Thus if the measured ^{51}Cr is, for example, 1/200th of the total amount of ^{51}Cr originally added, the technician knows that the volume of the red cells in the sample just drawn must also be 1/200th of the total cell volume in all the blood. Thus the total cell volume is precisely measured. The rate of cell survival can be determined by repeated samples over a period of a month or more.

Radiation doses from procedures such as these would typically be about 90 mrad in the blood itself, 1 rad in the spleen where blood cells concentrate, 200 mrad in the liver, and 10 mrad whole body. Assuming the effect of 1 rad approximately equal to 1 rem in this case, the amounts, except for the spleen dose, are within the range of naturally received radiation and no harmful effects would be expected. The localized dose to the spleen does not appear to be harmful either.

At one time gold-198 (^{198}Au) was used as the radioisotope through which to obtain liver scans. The isotope decays with a 2.7 day half-life and gives off an energetic β particle (0.96 MeV). Use of this isotope resulted in absorbed doses in the liver of as much as 6 rads from an administered dose of 150 μCi of the isotope. This is a higher than desirable dose, and in recent years the gold isotope has been replaced by use of technetium-99m. The m signifies that this is the metastable form of ^{99}Tc, that is, the one which is unstable and transforms quickly. It has a half-life of only 6 hours and emits a 143 keV γ radiation ideal for scanning purposes. Regular technetium-99 has a half-life of 200,000 years and is thus only weakly radioactive.

Technetium (Tc) is a fascinating element. It does not exist naturally at all. It is produced in nuclear reactors as one of the more abundant fission products or can be specially produced in a nuclear reactor by bombardment of natural molybdenum (Mo). In this method the absorption of neutrons in stable ^{98}Mo produces ^{99}Mo, which is a radioactive isotope with a half-life of 66 hours decaying to ^{99}Tc. We have pointed out that ^{99m}Tc has a half-life of only 6 hours, so once it has been separated chemically as pure technetium it disappears quickly and is essentially gone in about 3 days. This means that a hospital wishing to use it has to work very rapidly, particularly if the site where the chemical purification of the technetium occurs is some distance from the hospital. To simplify the problem the hospital usually orders the molybdenum-99 isotope, which acts as a "cow" from which the technetium-99m can be "milked" right at the hospital where it is needed. This works because the ^{99}Mo produced by bombardment is continually decaying to ^{99m}Tc, which builds to an equilibrium amount where its rate of decay just equals its rate of formation. It's sort of like a jar with a few green jelly beans mixed in with a lot of white ones. If white jelly beans turn to green just as fast as you remove green ones from the jar, you'll always have the same number of green jelly beans (technetium-99m) in the jar.

There is now a specialized business supplying ^{99}Mo to hospitals doing their own chemical separation of technetium for immediate diagnostic use in patients. This has led to a large increase in the use of ^{99m}Tc in view of its favorable radiation emission energy and its short half-life. Complex scans can be made without large total doses to the patient. There has been particularly brilliant success in the use of ^{99m}Tc for brain scans to determine the areas of tumors or damage to the circulatory system. This application has increasing competition from computerized tomography, but each method has advantages and both will continue to be used.

There are many other areas where radioactive isotopes are used for medical diagnosis. ^{14}C, ^{15}N, ^{42}K, ^{85}Sr, ^{75}Se, and ^{125}I are additional commonly used species. The rate of use has been increasing steadily since the middle 1950s and is still growing. It is estimated that in 1980 perhaps 1 in every 100 people in the United States had some form of diagnostic test involving radioisotopes. You will be an unusual person if by the time you die you will not have had one too. While the number of tests has been increasing, there has been strenuous effort to reduce the radiation exposure per test. Similarly there has been pressure to limit the use of radioisotopes to cases where the reward in information gained is clearly worth the small, long-deferred danger that may exist from the levels of radiation received.

RADIATION THERAPY

We now come to the aspect of radiation where the risk/reward relationship changes markedly. In most of this chapter we have been speaking of significant rewards, but at very low risk. In radiation therapy the reward is life itself, sometimes for only a few months or years, but often for much longer than that. The risks of side effects or delayed harm may be considerable, however. Doctors and patients have to weigh and judge these risks, but the alternatives may be grim and the rewards, even though for a short time, may be priceless.

It is perhaps paradoxical that radiation in relatively modest amounts may be instrumental in initiating cancer but in very large amounts can destroy the same cancer it induced. It is a fortunate fact that many types of cancer cells are more easily damaged or destroyed by radiation than are normal cells. Thus if X- or γ radiation can be concentrated as much as possible on the cancerous tissue itself, the cancer can be eliminated with only moderate side effects to essential normal tissue. For these treatments, however, very large total doses must be given, in some cases as high as 6000 to 7000 rads in a local area. These are cumulative doses and will be achieved through multiple treatments of perhaps 200 rads each on a regular basis over several weeks or months. The issue is survival, and the possibility of other damage years later can be overlooked, particularly in older patients.

In a typical radiation therapy treatment a careful map is made of the locations of the cancer tissue to be destroyed. Marks are made on the patient's body to help the radiation therapist aim the radiation beam at the exact location desired with proper calculations for the angle and the depth of the tumor. Doses are calculated to be the maximum tolerable by the healthy tissue, and usually the radiation is repeated on a daily or weekly schedule. Great care is taken to use well-focused beams to minimize radiation to other parts of the body, and lead protective shielding may also be used for the same purpose.

Much early radiation therapy was done using the radiation from ^{60}Co. This produces a 1.3 MeV γ and has a half-life of 5.3 years, so a purchased source could be used for a long time. It is difficult to focus such radiation as precisely as would be liked, and in recent years there has been increasing use of X-ray machines such as linear accelerators. In these devices a beam of high-energy electrons is focused on a target such as the metal tungsten. The tungsten is thus induced to emit X rays of a characteristic wavelength that can be focused very exactly. This is essentially the same system used in diagnostic X-ray machines, and the difference lies only in the higher X-ray energies produced. Electron beam energies as high as 35 million volts are available that permit high X-ray dose rates to the patient. Dose rates of 100 to 500 rads per minute are common, and since the typical maximum dose per treatment might be 200 rads, this leads to very short concentrated treatments.

While the great bulk of radiation therapy is carried out with X rays and γ rays, there are specialized or experimental uses of other radiation types as well. Neutron therapy, for example, has the advantage of extremely high penetrating power and is used at several major locations in the United States. High-energy electron beams can be used directly without conversion to X rays. They have relatively low-penetrating capability, however, and are used largely on surface cancers. Protons have the interesting characteristic of having a high-absorption rate just at the end of their range of penetration. They can be used effectively where the cancer is just the right distance below the skin surface and have been particularly useful on tumors of the pituitary gland, which is located just at the base of the brain. Even the use of pi-mesons is being explored at laboratories at Los Alamos, New Mexico, and at Stanford University in California, and the range of radiations used will undoubtedly expand further.

Radiation therapy has also been used for a number of noncancerous diseases, particularly skin problems. This usage has declined markedly in recent years, however, as more has been learned about the long-term effects of moderate radiation doses and as other safer treatments have been found.

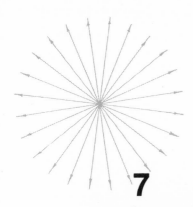

7

ELECTROMAGNETIC
RADIATION: MAN-MADE

THE DISCUSSION OF THE MEDICAL USES OF X RAYS and gamma (γ) rays brings us easily back to the electromagnetic spectrum first discussed in Chapter 1. It includes (in order of increasing wavelength) γ rays, X rays, ultraviolet light, visible light, infrared and heat, microwaves, and radio waves. We have seen how γ radiation is a result of radioactive decay and is present both as a natural part of our environment and as associated with nuclear power, nuclear weapons, and nuclear medicine. X rays are largely a creation of man, however, and we will turn our attention to them for a moment to learn a bit more about their origin and where, in addition to medical applications, we may encounter them.

X RAYS

X rays are generated when high-energy electrons strike a target of a heavy metal such as tungsten or gold or lead. This is the basis for the high-energy X-ray tubes used in medicine and in industry. Electrons are the basic units of electricity and as such, with proper circuitry, can be accelerated to very high energies in concentrated beams that can be focused on a tungsten target, for example. The high-energy electron beam striking the tungsten target also generates large amounts of heat, and sophisticated X-ray tubes may involve rapidly spinning targets so that the heat can be dissipated along a path on the target surface. Even so, very high temperatures are generated, and it is the very high-melting

81

temperature of tungsten that makes it capable of absorbing such punishment while emitting the very energetic X-ray beams desired.

In addition to their use in medicine, X rays are extensively used in industry, particularly for diagnostic testing of metal parts to ensure their freedom from flaws. Industrial X-ray machines have energies as high as 30 million electron volts (MeV) and can penetrate up to 2 foot sections of steel. Operations such as these must be carried out inside special protective buildings with thick concrete walls and labyrinth entries. Operation is done remotely in order to protect the personnel involved. With properly designed facilities there is no risk to people outside the building, and the only risks inside are from careless operation. All personnel are carefully trained and must wear monitor badges to check on any exposure that might be inadvertently received.

Probably all of us have spent many hours watching television, but few of us have recognized that television sets, particularly older models, have been sources of small amounts of X-ray exposure. The TV picture is formed by the scanning of an electron beam across the tube face, and we have seen that high-energy electrons striking particularly heavy elements are the creators of X rays. Voltages and materials are selected to minimize X-ray production, but small amounts can be emitted, particularly from poorly designed or out-of-adjustment sets. In the 1960s the International Commission for Radiological Protection (ICRP) recommended that X-ray emission from color TV sets should not exceed 0.5 milliroentgens per hour (mR/hr) at 5 centimeters (about 2 inches) from the surface of the receiver. Present U.S. regulations are stricter than that, and surface radiation is probably only 0.001 to 0.01 mR/hr in sets currently produced. At normal viewing distances exposures will be much less than this, and an average viewer's exposure in a year is at worst a tiny fraction of radiation received from natural sources or from medical X rays.

LIGHT AND HEAT

It is hard for us to recognize that light and heat are the same phenomenon as γ rays, X rays, microwaves, and radio waves. We tend to think of light and heat as benign and natural and the other wavelength radiations as somehow unnatural and potentially dangerous. Both reactions are wrong, but in opposite directions. Light and heat can have dangerous sides also, and the other wavelengths in rational amounts are as benign as light and frequently highly beneficial. We should somehow shelve our prejudices and see all of the wavelengths objectively.

Our purpose in this portion of the book is primarily to consider origins and sources of radiation, however, and we should return to that focus. The origins of visible light are all too familiar to us. The sun, electric light, flames and the combustion process, fluorescence both natural and artificially generated, and even the distant stars are all examples of light sources. We recognize these sources because of the incredibly beautiful perceptive capability of the combination of eye and brain. Many of these sources also generate frequencies higher or lower than visible light — ultraviolet for example on one side and infrared on the other. This unperceived light radiation can have significant effects as can be attested by the many beach goers who have received painful sunburns on what seemed to be a cloudy day. These burns were produced largely by the invisible ultraviolet portion of the light spectrum at wavelengths just below the visible range. It seems odd that the body has not developed a sensor to detect such damaging radiation, but it has not done so.

Thermal radiation (heat) is the portion of the electromagnetic spectrum of wavelength just beyond visible light and usually considered to include part of the so-called infrared light spectrum. It extends up to the region labeled "microwaves" and includes wavelengths from about 0.001 cm to perhaps 10 cm (about 4 inches). The heat we receive from a fire in the fireplace is almost all radiant energy, since the hot air generated all goes up the chimney. The energy of thermal radiation varies as the fourth power of the absolute temperature of the body doing the radiating. The absolute temperature scale has 0 at the coldest achievable temperature, where all molecular motion ceases. It may use degrees of the same size as Fahrenheit degrees and is then called the Rankine scale (in which ice freezes at 459.67°R), or it may use degrees of the same size as the Celsius or Centigrade scale and is then called the Kelvin scale (in which ice freezes at 273.15°K). These details are necessary to establish a qualitative feeling for radiant thermal energy. Let's stick to the Fahrenheit and Rankine scales because they are the most familiar in the United States, even though they are being replaced, and let's say that ice freezes at 460°R. If we take an object at the freezing temperature and double its temperature to 920°R (which is just 460°F), we will have increased its radiant energy by a factor of 16. If we double it again to 1840°R (1380°F), a temperature at which the object will have a dull red glow, the radiant energy will be 256 times that at the freezing point. Thus the amount of radiant energy goes up extremely rapidly as temperature increases. That is why the radiant energy from the sun, even when dissipated in the enormity of the solar system, is sufficient to warm us here on earth. The equilibrium temperature of the sun is 5776°K or 9720°R and it has a radius of 432,000 miles, so the radiant energy it emits is stupendous.

The ball of fire from a 20-kiloton nuclear weapon explosion has an instantaneous temperature of 300,000°K, far hotter than the sun, but in a sphere with a radius of only 50 feet. After about 1 second it has expanded to about 150 yards, and the temperature has dropped to just about that of the sun. It is these extremely high temperatures that generate the viciously destructive thermal radiation from nuclear weapons.

Anywhere man creates high temperatures he provides a source of thermal energy in the wavelength range defined as heat. An incandescent electric light bulb has a tungsten filament operating at about 2850°K to provide a source of light, but actually about 90% of its radiation is in the infrared or thermal range, so in that sense it emits more heat than light. Man himself emits radiant energy in the form of heat. When he emits more energy than he receives he feels cold, and when he receives as much or more than he emits he has a sensation of warmth. Standing in front of a cold window on a winter day will give a sensation of chill on the side of the individual facing the window, not because the "cold" is reaching the individual, but because the individual is radiating more heat to the window than he is getting back (i.e., he has a heat loss).

In general, however, people have an intuitive understanding of heat and how it behaves. We know enough to keep our distance from radiating sources such as fires or the white-hot metal produced in a steel mill or the molten lava from an active volcano. We became familiar as children with the ability to focus the thermal radiation from the sun with a magnifying glass so as to generate enough local heat to start a fire. We can see the glowing origin of thermal radiation, or we can sense its effect on our skin and hence avoid receiving too much. We know how to use it to cook food, heat our homes, or carry out industrial processes requiring heat. Except in rare circumstances thermal radiation is our servant and friend. Like light it lacks the mystery of the unfelt and unseen. In spite of this the toll of injury and death from thermal radiation undoubtedly is higher than from any other form of radiation.

We only have to move further out the scale of wavelengths, however, to enter another region of electromagnetic radiation that we use constantly but do not sense or understand. This is the area of microwaves and radio waves, which extend from the upper boundary of infrared at wavelengths of about a tenth of a centimeter (0.1 cm) to very low frequency radio waves with wavelengths up to 100 kilometers.

MICROWAVES

In more precise terminology the microwave region of the electromagnetic spectrum is defined as the region between 100 gigaherz (GHz) and 10 megaherz (MHz). In terms of wavelength this is between 0.003

meters and 30 meters. Longer wavelengths are considered radio waves. Microwaves are important because in this band of frequencies we include microwave transmission systems, radar, microwave ovens, television, and diathermy systems. They are also important because they cover a frequency range in which the energy they carry can be absorbed in the human body. At higher frequencies, in the infrared for example, the energy is absorbed in a very small surface layer or "skin depth." At lower frequencies in the radio range the radiation is reflected away from us without penetration. Microwaves, however, are absorbed, although the pattern of absorption varies with the particular frequency. The absorbed energy is largely evidenced by a heating effect, and if the energy density is high enough, actual increases in body temperature will occur. We will discuss the nature of the effects of absorption of microwaves in the body in Chapter 9. We will first, however, talk a little bit more about what they are, where they come from, and where and to what extent we encounter them.

Most of us own television sets and are familiar with the fact that there are two major television bands, ultra-high frequency (UHF) and very high frequency (VHF). The former includes the higher number channels from 14 to 82 and the latter contains channels 2 to 13. The higher frequency UHF (from 432 to 728 MHz) is probably the most significant contributor to our radiation environment, but VHF and FM broadcasting that are located in the frequency band from 54 to 216 MHz are also important. These broadcast systems usually generate their microwave signals by use of vertical antennas radiating out in all directions. Their intent is to have a signal wherever there is a television receiving antenna, so the radiation touches all of us. For those of us at an appreciable distance from the transmitter the signal is very weak, but for those few individuals close to the source antenna the microwave fields may be appreciable.

To understand the amount of this type of radiation we receive, we have to think again in a new set of units. For microwave exposures we speak of the power density of the field in terms of watts per square centimeter (W/cm^2). A watt is a measure of power with which we are familiar from our household light bulbs, which may range from 10 W to 150 W. You probably have a 60-, 75-, or 100-W bulb in your reading lamp. For a light bulb the power measurement is of the electric current consumed by the bulb, but it will serve to calibrate you on power levels. When we speak of W/cm^2 we are describing an amount of power passing through an area consisting of a square one centimeter on a side. It is thus an amount of power passing through a given area. One watt per square centimeter is quite a lot of power, so for microwaves we usually speak of microwatts (millionths of watts) per square centimeter ($\mu W/cm^2$) or milliwatts (thousandths of watts) per square centimeter (mW/cm²). At levels above 10 mW/cm² some body heating will be generated by the absorption of microwaves, so in the United States that is the limit of permissible radiation levels. In the USSR the recommended limits have been set much lower, and we will discuss this strange situation later.

The Federal Communications Commission has placed a maximum limit on the amount of power that can be radiated from a UHF-TV antenna. This limit is 5 million watts (5 megawatts or, as abbreviated, 5 MW), so the power density very close to such an antenna will be quite high. In a direct line from a 1-MW antenna the maximum permissible power density of 10 mW/cm² will be exceeded anywhere closer than 500 feet. This is on a direct horizontal line, however, and the antennas are almost always placed at high elevations to increase their range of coverage. At levels below the antenna height the power density drops off markedly.

There has been valid concern expressed about the situations in large cities where in some cases several UHF antennas have been placed on the top of tall buildings. Radiation levels on the ground will be low because of the elevated placement of the antenna, but if there are other equally high buildings close by, the radiation levels in the upper floors of these buildings may be considerable. For example, the high-rise building known as One Biscayne Tower in Miami, Florida, is located near an FM broadcast antenna. Microwave power densities have been measured at various floor levels in this building, and they increase from 7 microwatts per square centimeter (7 μW/cm²) at the 26th floor to 62 μW/cm² at the 34th floor, 97 μW/cm² on the 38th floor, and 148 μW/cm² on the roof. Measurements at the 102nd floor observatory in the Empire State Building in New York have indicated microwave levels of 30 μW/cm² near the window, and an FM microwave level of 200 μW/cm² has been measured on the roof of the Sears Building in Chicago. These measurements are all well below the 10 mW/cm² recommended limit and are not believed to be harmful even for continued exposure. The possibility of subtle, low-level effects cannot be totally ruled out, however.

Another way of looking at the all pervasive nature of TV and FM signals is to consider the number of people exposed to a given level of microwave radiation from a single transmitting source. It has been estimated, for example, that in Philadelphia about 1,300,000 people are exposed to microwave radiation of 1 μW/cm² or more by the signals from station WPHL, while in Washington 800,000 people are exposed to this level or more by the WDCA/WETA complex. For levels above 4 μW/cm² the number drops to 3000 in Philadelphia and 20,000 in Washington. These are extremely low levels and undoubtedly are harmless, but they are illustrative of the radiant world in which we live.

Satellite communication terminals on earth are another source of powerful microwaves. These stations use dishlike parabolic antennas up to 200 feet in diameter designed to send out concentric signals aligned perfectly parallel and beamed to outer space. They are like a carefully focused flashlight beam where the size of the beam many yards from the light is just the diameter of the lens from which it emerges. A small amount of scatter exists at the edges, but almost all of the power is maintained within the central beam. Since the target of these beams is a satellite in outer space, they are directed upwards and away from areas occupied by people. The beams are so powerful, however, that power density remains above 10 mW/cm² for distances up to miles from the source. The Goldstone Mars station, for example, has a power level of 450 kW (kilowatts), and the power density is only down to 10 mW/cm² at a distance of nearly 10,000 m (6.2 miles). Because of this, air traffic is prohibited in the air space directly above such stations, and there is a substantial area around the site to which access is not permitted except for working personnel in order to minimize irradiation from any side leakage. The safety of such systems, as with many systems in our industrialized world, depends on operational safeguards and careful use.

Radar is another powerful microwave source common in our environment. It includes the scanning systems designed to detect entry of enemy missiles or aircraft into our airspace and the aircraft control radar in use in every major airport and in the planes themselves. Radar is by nature a pulsed source of microwaves so that the signal reflected back from the object being tracked can be detected in the quiet times between pulses. The peak power density in each pulse is therefore much greater than the average power density, and we have the complex question of which is the more important level to consider. In addition many radars operate in a scanning mode as they sweep back and forth seeking a target. This further decreases the power density averaged over time at any given point, but it increases the area of space or terrain on which some radiation falls. Radar signals for the most part are aimed skyward and hence avoid populated areas. Scans close to the horizon may generate considerable signal strength at ground levels, however.

High-power radar installations are now relatively common. A 1972 study showed 229 unclassified pulsed radar sources with "effective radiated power" above 10 GW (10 billion watts). Here the term effective radiated power includes consideration of the additional power density achieved by focusing the beam. There are presumably additional very high-power systems that are classified for reasons of military security.

The use of high-power microwaves to transmit power from a solar power collector in outer space to a receiving station on earth is a fascinating future possibility. It is not without serious problems, however.

In one such concept a satellite solar power station (SSPS) weighing 25 million pounds with solar panels covering an area 7.5 miles long by 3 miles wide would be assembled in outer space. Its photovoltaic cells of silicon or gallium arsenide would convert solar energy to electricity nearly 10 times as effectively as on earth, because an orbit could be established that would keep the satellite in sunlight nearly 24 hours a day, and there would be no clouds or atmosphere to reduce the light intensity. The DC electric current produced at the satellite would be converted to microwaves with a frequency of 3000 MHz (10 cm wavelength) and beamed to earth from a loop antenna 1 km (0.6 miles) in diameter on the satellite. Output power would be between 7000 and 8000 MW, and useful power distributed from the receiving station on earth might be 5000 MW or about equal to the output of 4 large nuclear reactors.

Cost of transport of such a system to outer space would be at least $100/lb for a total of $2.5 billion. Add to this the cost of manufacture and assembly plus the cost of the ground receiving and distribution station and you are probably well over $10 billion for such a system. So the economics are only marginally attractive even considering that the fuel (solar energy) is free. A receiving antenna of about 7 km (4.3 miles) diameter would be required plus a safety zone of perhaps 1 mile all around to ensure against excessive microwave radiation to personnel. A receiving area of 6 to 7 miles in diameter would thus be required. Passage through this zone by air traffic would be inadvisable, since the power density in the central zone would be in the order of 10 mW/cm^2 and would be higher at higher altitudes. Some heating of the atmosphere above the receiving station would occur because of energy absorption in the air itself, and possibly interference with other microwave communication systems would have to be considered.

Such a system appears to be clearly within man's capability in the future. Its economics and the environmental problems associated with it will have to be carefully studied and evaluated. It will be another area where the benefits and risks of man's use of radiation must be carefully weighed.

In the case of microwaves for communication purposes, this weighing has already been done with positive results. The great advantages of microwaves for communication are their ability to be focused in straight lines from transmitter to receiver and the large number of individual signals that can be transmitted within a given range of frequencies. This is often stated by saying the "bandwidth" is large. There is also little loss of power with distance so that distances between relay stations can be relatively long, although they must be in line of sight. For these reasons much of the everyday communication for telephone and telegram service is by microwave links, and additional systems are coming into place rapidly for other industrial communication needs. The linking of computer systems by microwaves is becoming increasingly common, for example, and electric utility systems are beginning to use microwaves to relay information about the behavior of their widespread power networks. Fortunately the communication needs, unlike power transmission, can be accomplished at relatively low powers, and a microwave communication station may be emitting only 5 to 10 watts of power, so that power densities in the main beam right at the antenna are only 1 mW/cm² and fall off rapidly with distance. Thus although considerable increase in the number of microwave communication units in operation can be expected, they will not contribute seriously to our radiation environment.

We now come to a topic close to the heart of many housewives and to the stomachs of their husbands, namely microwave ovens. The capability of microwaves to penetrate and be absorbed in tissue is what makes them effective in cooking. Heat is generated uniformly *inside* the roast instead of having to penetrate from the outside as in a conventional oven. The threshold above which increases in temperature occur in tissue when subjected to microwaves is 10 mW/cm², and this has been set in the United States as the maximum level to which people can be subjected. Microwave levels inside a microwave oven must be substantially above this level and are typically perhaps 1 watt per square centimeter (1 W/cm²) for a home microwave oven with a power of 700 watts. The power is generated by a tube known as a magnetron. This generates the microwaves, and they are distributed into the oven by the equivalent of an antenna. Inside the oven the microwaves are reflected back and forth by the metal walls and are absorbed by the food being cooked. Glass, paper, ceramic, or high–melting point plastic dishes or containers are recommended since the microwaves pass readily through them, just as light passes through a glass window. Some plastics can melt just from the heat of the food being cooked, so they are not recommended. Metals such as aluminum, steel, or copper should not be used for containers in a microwave oven because the microwaves are reflected off them, and they may prevent the microwaves from entering the food or at least can cause uneven heating.

Since the levels of microwave energy inside the oven are high enough to cause physical burns, it is important that they are not allowed to escape from the oven at anything like full intensity. With a conventional electric or gas-heated oven one shouldn't put a hand into the hot oven with the power on. Similarly one uses caution with a microwave oven, but they are designed so that if the door is opened the power is automatically shut off, and there are independent systems to ensure that this takes place. In addition door seals are designed to minimize any leakage of microwaves into the kitchen. These seals cannot be perfect, however, and low levels of leakage can and do occur. All ovens produced since October 1971 have had to meet a set of radiation standards established by the Food and Drug Administration and therefore have, on the average, lower leakage rates than ovens manufactured prior to that date. The standard requires that leakage shall not exceed 5 mW/cm² at any point approximately 2 inches from the oven surface. At greater distances, of course, the levels will be substantially less. For example, if the maximum level of 5 mW/cm² exists at a point 2 inches from the oven, the level will have dropped to about 0.05 mW/cm² (50 μW/cm²) at 20 inches away.

An important distinction that must be kept in mind in considering radiation from microwave ovens is the difference between the terms "emission" and "exposure." The standard established for microwave ovens limits the level of emission from the oven to 5 mW/cm² at, to be precise, 5 cm (1.97 inches) from the oven surface. The *exposure* an individual receives from that same oven will depend on how much time he or she spends close to the oven while it is on. It is clearly unwise to spend a lot of time with your eye glued to the crack of a microwave oven door while the oven is on. It is equally clear that normal movements within the kitchen while the oven is on will result in exposure on the average to power levels of no more than a few microwatts per square centimeter for relatively short times.

The same frequencies assigned to microwave ovens (915 MHz and 2450 MHz) are used for microwave diathermy treatments. The term "diathermy" means the production of heat in body tissues by electric currents for purposes of medical healing. Normal body temperature is 98.6°F and may increase to as high as 105 to 106°F in the case of severe fevers. By the use of microwave diathermy, temperatures from 104° to 113° may be induced locally for treatment of muscle spasms, diseases of the joints, or some blood circulation problems. Radio wave therapy at 13.56 MHz is also used for the same purpose. Care must be taken to ensure very precise control of the temperatures and where they are generated. All devices for diathermy radiation now come under regulation by the U.S. Bureau of Radiological Health.

Some success has been achieved in recent years through the combination of diathermy and X rays or γ radiation for the treatment of cancer. Temperature alone may be helpful in some cases, but combinations of diathermy heating followed by or preceded by X-ray or γ radiation have also been useful. Experiments are continuing to reveal more about how temperature can make tumors less resistant to radiation and how temperature and radiation can be combined in more effective treatment paterns.

RADIATION FROM HIGH-VOLTAGE ELECTRIC POWER LINES

The transmission of electric power across the nation has in recent years been at increasingly high voltages. You are probably familiar with the great metal towers marching across the countryside carrying the clusters of cables delivering our power from the generating stations to the distribution centers. The voltage in an electrical system can be likened to the pressure in a water hose—the higher the voltage or pressure, the greater the amount of current or water delivered through a given size wire or hose. Our normal house voltage is 110 volts, but the high voltage transmission lines are at 138,000, 365,000, or even 765,000 volts. Higher voltages are under study. Most of these lines are alternating current or AC lines, like those in your home, which vary from positive to negative 60 times per second. This is known as 60 cycle current, or current at a frequency of 60 Herz (60 Hz), where the term Herz simply means cycles per second.

Transmission at very high voltages is desirable because in this way more power can be sent over a given right-of-way, and we need fewer "power corridors" crossing the country. Concern has been expressed that radiation from these very high-voltage lines can be harmful to people or animals living close by or inadvertently passing under the high-tension lines. Actually in this case we are not talking of radiation in the electromagnetic spectrum itself, but of time-varying changes in an electric field created by the current passing through the transmission line. Let's try to clarify the meaning of that last statement by going back to some first principles.

Any electrically charged object has an electric field around it that decreases with the square of the distance from the object. Thus at 1 foot from the object the field is 4 times as strong as it is at 2 feet and 16 times as strong as it is at 4 feet, so the field decreases rapidly with distance. Another charged object in the field will experience a force directly proportional to the strength of the electric field at that point, and the force will be an attraction or a repulsion depending on whether the 2 charges are of opposite sign (positive or negative) or the same. An uncharged object will not be affected at all by an electric field.

If our high-voltage transmission line is an AC line, the charge (related to the voltage) will be changing from positive to negative 60 times per second; so the direction and magnitude of the force in the electric field will be changing with the same frequency. This is what we mean when we speak of a "time-varying electric field" in the vicinity of a high-voltage line. These time-varying fields can cause tiny electric currents (from 10 to 100 microamperes) to flow in the body of a human or an animal standing directly under a 765-kV transmission line, for example. The question is whether these currents have any detrimental effects.

High-voltage direct current lines, sometimes identified as HVDC, are also beginning to be used at 765 kV and possibly eventually higher voltages. Electric fields will still exist around these lines, but they will be constant rather than time varying and for a given field strength do not generate nearly as large currents in humans or animals as do AC fields. We will not discuss them further here.

We must also mention that in addition to the electric fields around a high-voltage conductor there are also magnetic fields. Electricity and magnetism are connected because a moving electric charge generates a magnetic field around it. Hence a wire carrying electric current always has a magnetic field around it when the current is flowing, and the magnitude of the field is proportional to the current. Corollary to that, when a conductor moves through a magnetic field, current will flow in the conductor. A human being is not a very good conductor but is a conductor, and things like nerve action and other body functions depend on this ability to conduct. A human walking under a 765-kV line, therefore, will have miniscule electric currents induced by his motion through the magnetic field. As we will see later the magnetic field effects are considerably less than those from electric fields, however, and we will concentrate on the latter in future discussions.

In summary, therefore, the "radiation" effects near high-voltage transmission lines are not those of either particles or classical electromagnetic radiation such as light, heat, or radio waves. They are the separate effects of the electric and magnetic fields associated with the electrical charges and current flows in the lines. It is the combination of these fields, however, as they move away from the source of origin, which generates the true electromagnetic waves in the radio spectrum, so it is all part of the same radiant world.

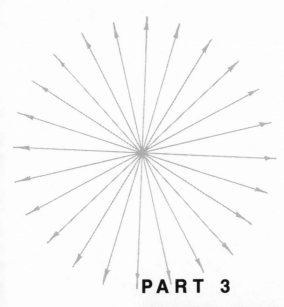

PART 3

EFFECTS OF RADIATION

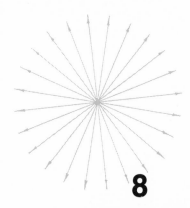

8

HOW AND HOW MUCH?

IF YOU HAVE COME THIS FAR you have learned what radiation is and where it comes from. You will have seen that it can be beneficial or harmful and that in general the greater the amount of radiation, the greater is the likelihood of harm. It is now time to understand the nature of this harm and how it occurs. This will help us understand also the controversy concerning possible effects of low levels of radiation such as we constantly encounter in our daily life.

ATOM'S EYE VIEW

We will look first at what happens in our bodies when a single particle, such as an alpha (α) or a beta (β) particle, or a single photon (the basic energy quantity) of γ or X-radiation whizzes into or through us. Remember that as you read this sentence you are being bombarded by the natural radioactivity from ^{3}H, ^{14}C, ^{40}K, and ^{87}Rb in your body; the radon and daughter products you have breathed in; and the natural cosmic radiation that has penetrated the atmosphere. Let us consider how these various particles or photons of energy interact with our own atoms and molecules.

To a poking finger the human body seems pretty dense and solid. To a probing α or β particle, because of their tiny size, most of human or animal tissue is empty space. The atoms making up our bodies are assembled into many different complex biological molecules, but to an α or β particle even these molecules have huge spaces between the atoms. For this reason and because of the nature of fundamental nuclear forces

95

an impinging charged particle hardly ever strikes an atomic nucleus directly. For example, in its passage through the interatomic space in our bodies, a swiftly moving β particle interacts with other electrons in orbit about a normal atom and can cause these electrons to be ejected from their orbit. This is the process known as ionization. Each interaction of this kind slows down the speeding β particle, until after a number of interactions it has lost so much energy that it can no longer cause further ionization. It then becomes absorbed as just another of the normal electrons that are part of the atoms of the body.

How far a β particle will penetrate in tissue depends upon its energy. A 5 MeV β will have a range of 2.2 cm (just under an inch), and a 1 MeV β will penetrate only 0.4 cm. The effects of external β radiation are therefore confined to the body surface, even where very high energy βs are involved. Protective clothing can provide almost complete shielding against external β radiation. It is for this reason that the greatest concern with β-emitting isotopes is for cases where they are incorporated in the body, by ingestion in food or water or by breathing into the lungs.

You will recall that a curie of radioactivity is that amount which undergoes 37 billion (3.7×10^{10}) disintegrations per second. A normal human body contains in the order of 14 mg (milligrams) ^{40}K, which is a naturally occurring β-emitting radioisotope. This amount of ^{40}K (about 100 pCi or picocuries) gives rise to about 4000 disintegrations per second in our bodies, so in the few seconds it has taken you to read this sentence your body has been bombarded internally by more than 10,000 1.3 MeV β particles from ^{40}K alone. The passage of a single particle through tissue is clearly not a very traumatic event, nor, so it seems, is the passage of thousands per second, since it is occurring to all of us all of the time. It is only when tens of thousands or millions or more of these events are occurring in relatively small areas that real damage to cells can occur.

Some particles are more damaging than β particles, however. Earlier we described α particles as helium nuclei with a double positive charge and a mass of 4 mass units, nearly 7500 times as heavy as a β particle. These heavy, highly charged particles cause large amounts of ionization of the atoms through which they pass, but because of their large size and the large number of interactions they create, they slow down rapidly and have short range. A 5 MeV α particle has a range of only 0.0037 cm in tissue, for example, as contrasted to 2.2 cm for the β particle discussed above. The damage produced by an α particle over its short range of travel is substantially greater than that of a β particle over the same distance. For that reason significant quantities of isotopes emitting α particles absorbed internally by breathing them into lungs or ingesting them in food or water can be more serious than similar quantities of isotopes emitting β particles. This is a concern with plutonium, for example. Even so we all have measurable quantities of α-emitting particles in us. Their source is the radon naturally emanating from the earth and depositing its most critical daughter product ^{210}Po in our lungs and in our food and drinking water. An average adult inhales 0.3 pCi/day ^{210}Pb (which decays slowly to ^{210}Po) and 0.07 pCi ^{210}Po itself, which is an α emitter. It is interesting and perhaps significant that someone who smokes 20 cigarettes per day increases his daily intake of ^{210}Po by more than 20 times, (from 0.07 pCi to 1.6 pCi).

Neutrons do not carry any electrical charge, and their interaction with matter is by direct collision with the nuclei of the atoms rather than by ionization. When a neutron strikes such a nucleus it may merely bounce off just the way one billiard ball bounces off another, or it may be absorbed into the nucleus causing a change in the nature of that nucleus together with emission of other radiation, usually a gamma (γ) ray or a β particle. In the bouncing or elastic collision the struck nucleus is knocked away, just as is the case with a billiard ball, and the neutron loses part of its energy, just as much, in fact, as is gained by the struck nucleus. These events are not as common as the ionizing events that occur with α and β particles, so neutrons have much greater penetrating power. A high-energy neutron can easily pass entirely through the body, for example. The biological effects of neutrons are complicated by the fact that when they are absorbed in another nucleus, either β particles or γ rays and in a few cases α particles are emitted as by-products. Some of the damage caused by neutrons is thus really caused by secondary β or γ radiation.

Gamma radiation and X rays can also produce ionization effects in material through which they travel. Although we have described them earlier in terms of their wave characteristics, they can also be considered as made up of pulses of energy known as photons. These photons of energy act as if they were particles without mass but capable of interacting with the electrons of the material through which they pass. One rad of 1 MeV X rays is equivalent to the passage of 2.2 billion photons per square centimeter of tissue, for example, and it is the interactive capability of these photons that causes ionization damage effects similar to the passage of α and β particles. Gamma and X rays have less interaction per length of path traveled, and hence have much greater penetrating power. This is what makes them so useful as diagnostic tools.

From the above discussions we can see that for the same initial energy different radiations have quite different effects. Alpha particles do the greatest damage per unit of distance traveled, but they only go very short distances. Neutrons have long-range and multiple secondary effects and are also very damaging. Beta radiation, X rays, and γ rays cause about the same amount of damage. The use of the rem unit tries to take this into account, so 1 rem of any of the different radiation types has the same total damage effect.

Karl Z. Morgan, one of the nation's leading health physicists, has pointed out that there are 4 possible responses of body cells to radiation impinging on them. These are:

1. The radiation can pass through without causing any effect.
2. The radiation can kill the cell or damage it so severely that it can't reproduce.
3. The radiation damages the cell, but the cell can repair the damage and continue to function normally.
4. The cell lives, but is damaged in such a way that it duplicates itself in a damaged or perturbed form. This can become a malignancy or cancer.

Before seeing how these responses relate to radiation dose levels, we need to examine also the mechanism by which the lower portion of the electromagnetic spectrum such as light, microwaves, and radio waves can cause biological damage. These longer wavelength, lower energy radiations do not displace the electrons of the material through which they pass. They are thus nonionizing, and you will often see them referred to as nonionizing radiation. They do transfer energy, however, in the form of increased vibratory action of the atoms and molecules through which they pass. This is what we sense as heat, so in simple terms nonionizing radiation merely heats up the material through which it passes. You can feel the warmth of the sun on your skin on a summer day and you can cook a roast in a microwave oven. These are simple manifestations of the heating effect of nonionizing electromagnetic radiation.

If this were the only effect of nonionizing radiation, the situation would be simple. When the radiation ceases, the temperature returns to normal. The only question would be whether the heating effects during the time the radiation was on were severe enough to be harmful. This may be the case, but there is some belief that more subtle effects than this exist also. This belief is based on the possibility that since the radiation is electromagnetic in nature, certain frequencies may disrupt or alter the tiny electric signals that govern the functions of our mind and our nerve and sensory systems. Bear in mind that these effects are far from proven, but the possibility and the evidence presented has given rise to major areas of controversy.

HIGH-LEVEL RADIATION

At this point we will return to the ionizing portion of the radiation spectrum to see what the physical effects are from large and small doses. The former have been made abundantly clear by the effects of nuclear weapons, by a number of severe accidental exposures, and by medical misuse of X rays before their side effects were known. The latter are still the subject of much controversy, which we will try to examine from both sides.

HIROSHIMA AND NAGASAKI. The physical effects of high levels of radiation have been most dramatically and horribly displayed by the bombing of Hiroshima and Nagasaki in 1945. While most (85 to 95%) of the 140,000 casualties were the result of blast and heat, radiation sickness contributed to the death of many of the badly injured and took a toll of its own as well. Long-term effects of the large radiation doses are still appearing, and careful follow-up study on the affected populations has taught us much about the effects of a single exposure to very large doses of mixed radiation.

There are four aspects of this experience that must be considered. First is the direct immediate effect of very high levels of radiation that cause a multitude of symptoms lumped under the term "radiation sickness" and result in death or recovery within a few weeks or months. Second are the long-term effects that can induce one or another of various forms of cancer over a period of 5 to 50 years or more. Third are effects on foetuses being carried by their mothers at the time of the explosions. These children showed unique symptoms related to the high susceptibility of the human foetus to radiation delivered early in its development. Fourth are the genetic effects induced by these large radiation doses. These same classes of damage can be used in much of the later discussion of radiation effects also, because they are four fundamental and distinct areas in which damage can occur.

The radiation levels that in a single exposure will produce acute radiation sickness are now pretty well known. A radiation dose above 600 roentgens (R) will result in death to almost all people receiving it, while about half the people who receive 400 R will die from it. Doses from 100 to 300 R are generally not fatal but will cause severe illness, while doses under 25 R will not generate obvious symptoms. The typical pattern of radiation sickness at Hiroshima and Nagasaki in someone who had received about 400 R would be initial nausea and vomiting about 1 to 2 hours after exposure followed by perhaps a week or 10 days without obvious symptoms. The higher the dose, the shorter would be the respite before violent symptoms again occurred. By about the middle of the second week after exposure to 400 R, symptoms would reoccur beginning with loss of hair followed by fever, inflammation of nose and throat, the appearance of blood spots under the skin, bleeding in the mouth and urinary tract, ulcer formation and accompanying infection. All this is accompanied by a major decrease in white blood cell count. These white blood cells are the body's infection fighters, and their destruction by radiation renders the body easy prey to infection in the sores created by other cell damage. Death can occur anywhere up to 3 to 4 months after initial exposure, but patients who survive as long as 4 months usually recover apparently completely except for possible long-term cancer complications, which we will discuss next.

It is now recognized that one of the effects of ionizing radiation is the generation of cell damage that can lead to development of various forms of cancer later in life. It is important to emphasize the phrase "can lead," because most of the time cancer does not result, only sometimes, and the frequency, as is reasonable, increases with the amount of radiation. This is well illustrated in the history of deaths from leukemia, a form of cancer related to the formation in bone marrow of excessive amounts of white blood cells relative to red, in survivors of the Hiroshima and Nagasaki bombings. After the signing of the peace treaty between Japan and the United States at the end of World War II the two countries mounted a major effort to understand the true nature of the aftereffects of the two nuclear explosions. Through the efforts of the Atomic Bomb Casualty Commission (ABCC), careful measurements were made to determine the probable radiation doses received by large numbers of people whose locations were exactly known at the time of the explosions. A life-span study was initiated to follow a selected group of about 21,000 men and women whose dose history could be well authenticated. The medical history of these individuals has been closely followed over the intervening years and has been compared with the average histories of similar Japanese not exposed to bomb radiation. Table 8.1 shows the results for the 25-year period between 1950 and 1974 for people from Hiroshima only.

It is clear from these data that the number of leukemia cases is

TABLE 8.1.   **Leukemia incidence in Hiroshima survivors**

	Range of dose (rads)				
	1–9	10–49	50–99	100–199	Above 200
Approximate number of people in group	11,600	9100	2280	1430	1300
Number of cases of leukemia	14	18	7	13	30
Number of cases expected in same size group receiving no radiation	25.5	19.5	4.9	3.1	2.8

substantially larger than normal in the population receiving radiation doses above 50 rads and that the extra number of cases above normal increases at increasing radiation doses. Not generally realized, however, is that the absolute number of cases is relatively small. In a group of about 5000 people receiving above 50 rads, only 39 cases above normal Japanese expectancy of 11 were observed. Looked at one way this is nearly 4 times normal. Looked at another way it is a 0.8% rate of additional occurrence over a 25-year span or about 8 cases per 1000 people receiving these very large doses of radiation. Additional individuals have presumably died of leukemia since 1974, but up to that time there was no evidence that the rate of occurrence was increasing.

The figures for people from Nagasaki show an increase in occurrence of leukemia above normal only for radiation doses above 100 rads. Out of 2400 people studied who had received above 100 rads, 15 extra deaths from leukemia were observed in the period between 1950 and 1974. In a group of about 4300 people studied who had received from 10 to 100 rads, no extra deaths were observed. It was initially thought that the difference in the statistics between the two cities might have been because the weapon used at Hiroshima was a gun-type bomb based on ^{235}U while the one at Nagasaki was an implosion-type weapon using ^{239}Pu. The radiation from the Hiroshima weapon was believed to have a substantially higher yield of neutrons relative to γ rays than did the Nagasaki weapon, and this was felt to indicate that neutron irradiation has a somewhat higher likelihood of inducing leukemia than does γ radiation at comparable levels. However, recalculations in 1980 of the radiation intensities from the explosions at both cities resulted in lower estimates of neutron radiation levels at both cities and higher estimates of γ radiation at Hiroshima. These findings appear to make the data between the two cities more consistent if the leukemia cases are related mostly to γ ray levels. They indicate also that the body can tolerate slightly higher levels of γ radiation than originally thought. The role of neutrons in cancer initiation is still unclear, however, and further analyses are being carried out. The quantitative conclusions concerning increased susceptibility to leukemia initiation by levels above 50 rads are in general still valid.

Other forms of cancer have also shown an increase in rate of occurrence in survivors of Hiroshima and Nagasaki who received more than a 10 rad radiation dose. For women receiving above 100 rads there was an increased incidence of breast cancer of about 4 to 5 cases per 1000 women studied. Between 10 and 100 rads the extra incidence was about 2 cases per 1000 women examined. The incidence was highest in women who were between 10 and 19 years old at the time of the irradiation. Lung cancer also showed an increase of about 2 to 3 deaths above normal for every 1000 individuals studied over the 22-year period from 1950 to 1972. In actual count there were 100 lung cancer deaths among the individuals known to have received above 10 rads, whereas 78 would have been expected in any similar group receiving less or no radiation. Comparing deaths from all types of cancer including leukemia in Hiroshima and Nagasaki in people who received 10 rads or higher radiation dose, in the period from 1950 to 1972 there were 1159 deaths as compared to 920 that would have been expected in a normal Japanese population of the same size. Thus 239 extra deaths over the 22-year period can be attributed to the cancer-inducing effects of the radiation from the explosions. Looked at another way, where a cancer death rate of 4.4% would have been normally expected, a death rate of 5.5% occurred in the 21,000 people receiving these very large doses of radiation. It is clear evidence that single high-intensity radiation doses above 10 rads will increase to a small extent the likelihood of subsequent death from cancer. It does not, however, show that even these large doses of radiation are more dramatic in their effect on cancer induction than cigarette smoking, air pollution, or chemical carcinogens. A reliable study in the United States in 1975 showed that the death rate from lung cancer in men 35 to 84 years old was increased by a factor of 8.6 if they smoked 10 to 19 cigarettes per day. The increase in cancer deaths in those receiving above 10 rads at Hiroshima and Nagasaki was by a factor of 1.3.

The human embryo or foetus, particularly in the period of a few weeks after conception to perhaps 4 months into the pregnancy, appears to be relatively sensitive to radiation damage. For this reason a number of studies have focused on the fate of those developing humans who were in their mother's uterus at the time the bombs went off. Where the mothers received sufficient radiation to show active symptoms of radiation sickness the effects on the baby were profound. One study of 30 such women showed that 7 of the foetuses died in the uterus; 6 died at birth or in the first year afterward; and of the 17 that survived, 4 were mentally retarded. The toll was grim. It was also apparent that the forming brain could be damaged, because in those children irradiated in the foetal stage 3 to 13 weeks into pregnancy there was a definitely higher than normal occurrence of microcephaly, a smaller than normal head with a pointed forehead and usually accompanied by mental retardation. Other forms of mental retardation were also observed particularly where the radiation dose was above 200 rads and occurred in the early weeks of pregnancy.

The fourth area of concern is genetic effects. There has been an often held belief that the children of survivors who received significant amounts of radiation at Hiroshima and Nagasaki have shown abnormal characteristics or shortened lives because of genetic damage to their parents. This belief is just not true, as shown by two careful and detailed studies. The first study covered a 17-year period of medical history for 20,000 children born to parents where one or both had been less than 2 kilometers (about 1¼ miles) from the central point in the city where the bomb was exploded. This history was matched with a second group of about 16,500 children, neither of whose parents had been closer than 2 kilometers from the explosion center but either or both had been between 2 and 2.5 kilometers from that point. Both these groups were then compared with a third group of about 17,000 children, neither of whose parents had been in Hiroshima or Nagsaki at the time of the explosion. There was no difference in death rates over the first 17 years for the children of any of these groups.

The second study looked for chromosome abnormalities in 2885 children whose parents (either one or both) had received more than 1 rad from the A-bomb at either city. The chromosomes are the tiny threadlike bodies found in the nucleus of cells, and they contain the complex DNA molecules determining hereditary characteristics. Damage to the chromosome structure can result in birth defects or hereditary changes. The abnormalities found were no different in type or number than those seen in a group of 56,000 otherwise similar children whose parents had received no radiation or less than 1 rad. From these studies and those discussed previously it is clear that if irradiation occurs during a mother's pregnancy, there is a real likelihood of death or damage to the child as yet unborn. A child conceived after the parents have received even substantial doses of radiation, however, appears to have the same chance for a normal healthy life as a child born to parents who have not been irradiated.

The nuclear explosions at Hiroshima and Nagasaki are the only cases where a very large number of people have received massive doses of ionizing radiation in a single pulse. The history and consequences of these events therefore spread before us the entire spectrum of damage such radiation can cause—from agonizing death at one extreme to no discernible effect at the other. The study of these catastrophic events is therefore enormously important to us in understanding the effects of radiation and in particular in helping determine whether there is a level of radiation that is truly harmless. Before embarking on the difficult task of identifying whether low levels of radiation are harmful, we need to review a number of other cases of accidental overexposure to radiation to see what we can learn from them in terms of their avoidance and in terms of the insight they give on the effects of various levels and types of radiation. Some of the cases are related to nuclear weapons or reactor development and some are from mistakes made in early use of medical radiation therapy. While we learn from them what large amounts of radiation can do, we cannot escape their fascination as human drama also.

LUCKY DRAGON CASE. One of the most dramatic incidents is the case of the *Lucky Dragon.* The ironically named *Lucky Dragon* was a Japanese fishing boat engaged in tuna fishing about 120 miles east of Bikini atoll in the Pacific Ocean in the very early morning of 1 March 1954. On board was a crew of 23 young Japanese fishermen between the ages of 19 and 39. About 3:50 A.M. they were puzzled by the sight of a great red glow in the west, followed shortly afterwards by a dull boom. Three hours later a strange white ash began to fall on their vessel, and it continued for about 4½ hours and then stopped. They gave up fishing shortly afterwards to return to their home port of Yaizu, but they did not reach there until 14 March, almost two weeks later, still not knowing that the ash was radioactive fallout from a 17 megaton thermonuclear weapon test (the Bravo test) detonated at Bikini.

The ash stuck to clothes, hair, and exposed skin and coated the boat with a film heavy enough to show footprints when walked upon. One of the fishermen was curious enough to collect some of the ash in a container and place it by his bunk. Within a few days nearly all the fishermen developed headaches, a feeling of fatigue, and loss of appetite. Eight of them experienced nausea and vomiting, and nearly all experienced severe eye irritation. Within a few days lesions or sores began appearing on the areas of skin directly exposed to the ash.

Even after their return to port recognition was slow in dawning concerning the real cause of their sickness. By 28 March, however, nearly a month after their exposure, all 23 had been hospitalized in Tokyo and were undergoing intensive treatment and care. One of the men died in September, nearly 7 months after the initial radiation. The rest gradually recovered, and all had received their final discharge from the hospital by May of 1955, more than a year after the white ash first fell on them. Of the 22 who survived, one died in 1975; but the cause of his death could not be clearly linked to his radiation exposure. The remaining 21 were still living in 1979 and were in apparently normal health at that time.

When the nature of the problem was first recognized by authorities on shore, measurements for radiation levels were made on various parts of the boat and on samples of the ash the crew had collected. From these measurements and knowledge of the decay rates of the isotopes found, it was possible to estimate the levels of dose received by the crew. Attempts were made to do this individually for each member of the crew based on where he slept and what his job was on the boat. Doses received in the first 24 hours were estimated at 100 to 150 rads external body radiation for the crew member receiving the lowest dose and 420 to 500 rads for the individual receiving the maximum dose. Over the whole 2 weeks of exposure while on the boat returning to port estimates ranged from 170 to 200 rads for the lowest dose to 660 to 690 for the highest dose. The man who received this largest estimated dose survived, and the man who died probably received 510 to 590 rads. Amounts ingested in food and into the lungs may have been significantly different in the two individuals, however.

All of the surviving fishermen who were not married at the time of the accident were subsequently married and had normal children, confirming the evidence from Hiroshima and Nagasaki that even substantial doses of radiation do not impair the genetic traits of future offspring. Thus the case of the *Lucky Dragon,* while a tragic event, showed that complete recovery and normal life expectancy is possible even after severe radiation dose extending over a 2-week period. This is an important benchmark in considering the effects of lower doses.

RONGELAP, AILINGNAE, AND UTIRIK. The crew of the *Lucky Dragon* were not the only people to suffer from the fallout from the Bravo test at Bikini that fateful 1 March. The tiny Marshall Island group contains a total of 34 islands including the weapons test islands of Bikini and Eniwetok. To the east of Bikini and at distances of 100, 110, and 300 miles respectively lie the islands of Rongelap, Ailingnae, and Utirik. Like the unlucky *Lucky Dragon,* they lay directly in the path of the windblown fallout cloud from Bravo test. There were 67 people on Rongelap, the closest island, and the fallout there was described as a snowlike ash. Farther away at Ailingnae, where 19 people lived, it was more like a fine mist; at Utirik (population 158), 300 miles from Bikini, the fallout was so fine as to be invisible. Estimates of average external γ radiation dose received are 175 rads at Rongelap, 69 rads at Ailingnae, and 14 rads at Utirik. Consistent with these dose levels were transient nausea and vomiting at Rongelap and to a lesser degree at Ailingnae. No such symptoms were observed at Utirik. Widespread skin burns were also noted in the closer-in islands along with loss of hair and lowering of the white blood count. No deaths occurred, however, and all those injured appeared to recover. Radiochemical tests of urine specimens soon after the incident showed the presence of radioactive isotopes of strontium, barium, and iodine, and particularly the latter appears to have resulted in residual harm.

Doses to the thyroids of adults at Rongelap have been estimated at 335 rads, and for children the estimates are as high as 700 to 1400 rads because of the smaller size of a child's thyroid. Within 10 years effects began to appear, including growth retardation in some of the children. Perhaps the most significant effect has been the appearance of nodules or lumps on the thyroid plus, in some cases, apparent decrease in the effectiveness of the thyroid's production of normal level of hormones. These hormones are essential in the regulation of body growth and the rate at which food is converted to energy. A 1979 report showed the appearance of thyroid nodules in 31 out of the 86 people exposed at Rongelap and Ailingnae and 15 out of 158 people at Utirik. A similar group of 437 people from other islands not exposed to radiation showed only 29 nodules. The percentage at Utirik is thus only slightly higher than for the unexposed group, but at the closer-in islands of Rongelap and Ailingnae it is much higher (36% versus 6.6%). It is highly probable that a combination of ^{131}I and the shorter lived isotopes ^{132}I, ^{133}I, and ^{135}I was the cause of these thyroid abnormalities. The significance of iodine as a radiation hazard in fallout is thus made clear, but it must be kept in mind that the levels in these cases were quite high. It is also well to recognize that nodules and low-function levels in thyroids such as these are treatable by hormone medication.

The islanders exposed to radiation were treated, and the decision was made to move them from their homes to other islands. They have instituted a major lawsuit against the U.S. Government to obtain compensation for their injuries and the trauma and disruption in their lives caused by Bravo test. Their story is not over, but we will have to leave it at this point.

Y-12 INCIDENT. The only saving grace of accidents is that they frequently teach us things we otherwise might not learn. This has been true of a serious accident that occurred in June 1958 at a nuclear facility called the Y-12 plant in Oak Ridge, Tennessee. This plant was one at which early work on separation of uranium isotopes was carried out using an electromagnetic method. The accident occurred when workers inadvertently added enriched uranium to a drum in such a way as to create a critical mass. As we described earlier, when a critical mass is achieved or exceeded there is an enormous multiplication of fission events and a huge release of energy and radiation. When purposely confined this creates a nuclear bomb explosion, but when unconfined it can rapidly blow itself apart and die out. This is what happened at Oak Ridge. In the instant burst of radiation that occurred, 8 men working in the area received substantial doses of γ rays and neutrons. Five of the men received whole-body doses of 236 to 365 rads, and the other 3 were estimated to have received below 70 rads.

Here was a case where immediate medical attention could be focused on the exposed individuals and the course of their reactions could be observed carefully. The 5 most seriously exposed men were hospitalized immediately and the 3 others within 2 days of the accident. These latter 3 had received an estimated dose of 22.5 to 68.5 rads. They showed no symptoms of "radiation sickness" and were discharged a week later. The 5 most seriously exposed individuals developed nausea and vomiting beginning within a day or so and lasting from 1 to 5 days. Their white blood count dropped rapidly, and after 17 days all suffered hair loss and other minor symptoms. After 44 days in the hospital all were discharged, and although reporting muscle and joint soreness, a tendency to tire easily, and some symptoms of nervousness and depression, all regained apparently normal health within about 6 months.

The men were all back at work 4 months after the accident, and their medical history since that time appears to have been relatively normal. Two of them have since died. Of these, 1 died at age 54 from lung cancer and the other at age 74 from a stroke. The lung cancer victim had been a hard-rock miner earlier in his career with 12 years of underground mining, and it seems probable that this activity rather than the radiation accident may have been the cause of his problem. The other 6 men were still living in 1979, 21 years after the accident, without medical symptoms clearly traceable to the radiation dose. A psychological examination in 1962, however, showed that fear of possible effects, genetic or otherwise, had colored their view of life and in 2 cases caused mild mental depression. The fear of radiation effects may thus have caused more long-term harm than the radiation itself, a point important to all of us.

SAFETY SYSTEM FAILS. The price of carelessness in maintenance of safety interlock systems was demonstrated in unfortunate fashion in the Pittsburgh, Pennsylvania, area in 1967. The accident involved a machine called a Van de Graaf generator producing an electron beam focused on a gold target from which emerged high-energy X rays. These were in turn focused on a beryllium target to generate neutrons for use in a complex chemical analysis, but it is the X-ray beam itself with which our story is involved. Entry into the chamber where the X rays were generated required opening a control panel, removing a special key, and then unlocking in succession the doors to a safety tunnel and the target room. Any one of these four actions was supposed to automatically turn off the electron beam and thus stop the emission of X rays. The system failed to work, however, and unbeknownst to the 3 maintenance men who had been asked to enter to make repairs on the cooling system for the generator, the X-ray beam was still on when they entered the room. It remained on while they worked.

The first man carried out the actual repairs while his partner stood behind him and handed him the needed tools. The third man was only present, fortunately for him, for part of the 20 minutes it took to do the job. Within about 30 minutes after completing the job the first worker felt nauseated and began to vomit. He reported to the medical department who first thought he had intestinal flu, but when his partner showed up with similar symptoms somewhat later a serious radiation exposure problem was recognized.

The film badges for all these men were immediately checked and the medical department's suspicions were verified. All 3 men were rushed to the hospital and treatment was begun. Later tests showed that the lead man had received approximately 600 rads whole-body radiation plus 5900 rads to his hands and 2700 rads to his feet. It was the combination of very high overall dose coupled to enormous local doses to hands and feet that made his case particularly difficult and tragic. His immediate partner received about 300 rads whole-body, but since he had stayed behind the leader did not receive significant extra local dose. The third man received only about 100 rads, and after 17 days in the hospital was released as essentially back to normal.

Both of the other men went through the by then well-recognized pattern of very large drop in the white blood count and the accompanying high risk of infection. The man who had received 300 rads gradually recovered with the help of a blood platelet transfusion about a month after the exposure. Forty-three days after the accident he was released from the isolated special care area in which he had been treated.

The man who had received the major dose was in much more critical condition. In addition to the drastic lowering of white blood count and other changes in his blood chemistry, the high-dose levels to hands and feet led to gross blisters, loss of skin and flesh, infection, gangrene, and eventual amputation of portions of both arms and legs. He did not die, and with the aid of artificial limbs and hands he has learned to care for himself and lead a useful life, including serving as a town commissioner.

A reading of the detailed medical history of this case shows that by 1967 doctors understood the typical pattern of severe radiation sickness and had developed a partially standardized mode of treatment. Central to this was a recognition of the expected changes in blood chemistry and the body's inability to fight infection. The method of care therefore involved isolation of the patient in a sterile environment into which entry was permitted only to authorized medical personnel in clean clothes, and where inlet air was carefully filtered to reduce environmental contamination. Antibiotic drugs were used to fight infection, and transfusions and bone marrow grafts were used to help regenerate healthy blood. Only through this skilled intensive care in specialized facilities was the life of certainly 1 and probably 2 of these accident victims saved. In a nuclear weapons attack local facilities for such care would be destroyed, doctors would be unavailable, and the facilities and medical skills in the region surrounding the devastated area would be completely swamped with serious cases they would be unable to treat adequately. This is why the lessons learned from small but serious radiation accidents should be more than medical ones.

WE ADD IT UP. There have been many other radiation accidents over the years, mostly involving misuse or unrecognized handling of radiation sources such as ^{60}Co or ^{192}Ir capsules for medical or industrial radiography. There has even been a case where a divorced father tried to kill or maim his 13-year-old son by deliberate close exposure to a 1 curie source of cesium-137. All of these cases have increased medical understanding of the clinical symptoms caused by high levels of radiation and how radiation sickness can best be treated. They have shown the intense local destructiveness of exposures up to thousands of rem where small areas of the body were exposed and destroyed yet where the patient lived. They have shown that whole-body exposure below 10 to 50 rems produces no apparent symptoms, that from 100 to 300 rems will cause severe symptoms but will usually not result in death, that exposure above 300 rems is increasingly likely to cause death, but that good medical care can conserve life even up to 600 rems. The data supporting these broad conclusions are clear and generally agreed upon. It is the question of the effects of exposure below 10 to 50 rems and down to very low levels that has been the subject of much bitter debate. In the following pages we will try to illuminate this controversy so that you may at least understand its nature and the basis for the opposing views.

LOW-LEVEL RADIATION EFFECTS

You might think it would be easy for scientists and doctors to carry out experiments or make observations telling us exactly what the effects of exposures to various levels of radiation would be. There are many reasons why this is not so, and we have to understand those reasons before we can understand the controversy and approach the truth. The first problem is that the effects, if they exist at all, are very small and occur in random fashion. Leukemia, for example, a cancer of the blood cells, is one of the forms most readily triggered by radiation. Yet in the 25-year period from 1950 to 1974 in a sample of about 20,000 people receiving 1 to 49 rads at the time of the explosion at Hiroshima, there were fewer deaths (32) from leukemia than would have been expected in a normal unirradiated group of the same size (45 expected). In a group of about 2200 people receiving an estimated 50 to 99 rads there were 7 leukemia deaths in the same 25-year period versus 5 that would have been expected in a normal population. In the somewhat smaller group studied (about 1400 people) who received 100 to 199 rads there were 13 leukemia deaths versus 3.2 expected. This is a fourfold increase, and the effect of radiation is beginning to emerge clearly. For those receiving 200 rads or more the risk of leukemia increased with increasing dose, but even in the group of about 400 survivors who received 400 rads or more there were only 12 deaths from leukemia in that 25-year period where perhaps 1 would have been expected in a similar group not receiving radiation. This is a twelvefold increase in probability, but it is still only a 3% death rate from leukemia.

From this background our problem is to determine whether there is a leukemia risk from radiation at levels of 10 rads and below or from, let's say, 1 rad per year for 40 years. Common sense tells us from the Hiroshima data that the effect will either be very small or nonexistent, but common sense can't tell us surely which is correct. This is what is called the problem of extrapolation. How do we tell the effects of low levels of radiation on very large numbers of people when our only clear data are from high levels of radiation on relatively small numbers of people and even these involve small effects?

Another complication is the absence of a medical way of telling what caused a particular case of cancer. For any million babies born today we can expect that 164,000 of them will die of cancer. Known or suspected causes of cancer include toxic chemicals, viruses, inherited genetic defects, and many others in addition to radiation. In a normal life an individual will receive on the average 100 mrems/yr of mixed natural radiation, and we lack ways of knowing whether or not radiation is a causal factor in any of the cancer deaths.

A direct experimental approach on man is hardly possible. We cannot ask thousands of volunteers to submit to various amounts of radiation in order to follow scientifically any medical consequences that may result. The Hiroshima/Nagasaki weapons explosions, accidental radiation releases, and careful studies of occupational exposures are thus the only basis for direct study of effects on humans. These are often complicated by inaccuracies in knowledge of doses received, lack of knowledge of other factors influencing the data, and often by small sample size. Experiments on animals are another valuable source of information, but animals may be more or less sensitive than man, and we have no sure way of telling the degree of this difference.

Finally, and unfortunately, the question of radiation effects has often been beclouded by emotional factors. The elimination of nuclear weapons and nuclear power has become an article of faith to many laymen, doctors, and scientists, and the zeal to prove the harmful nature of radiation has sometimes colored their objectivity. Similarly at the other extreme the supporters of nuclear power have been prone to gloss over or even hide data detrimental to their thesis that nuclear power is a desirable and beneficial force. Our task is to clear away the cataracts that dim the vision of both sides and try to see the results clearly. We will start with the low-level effects of ionizing radiation and then look at nonionizing radiation such as microwaves and transmission line radiation.

IONIZING RADIATION. The two bodies that more than any others are the conscience of science and engineering are the National Academy of Science and its offspring the National Academy of Engineering. The National Academy of Science was established in 1863 by Congress as a private, nonprofit, self-governing membership corporation. Its function is, whenever called upon by a department of the government, to "investigate, examine, experiment and report upon any subject of science or art." Members are elected after nomination by the various sections, and membership is one of the highest honors in American science or engineering. In 1916 the National Academy of Science established a National Research Council (NRC) to become its principal operating agency, a function NRC now fulfills for the National Academy of Engineering and the Institute of Medicine also.

Through the National Research Council Advisory Committee on the Biological Effects of Ionizing Radiation (frequently referred to as the BEIR Committee) a series of studies were undertaken, starting in the summer of 1970, to establish a scientific basis for setting irradiation standards. These studies have broadened to include an assessment of the consequences of exposure to low levels of ionizing radiation (X rays, γ rays, neutrons, and α and β particles). The committee consisted of 23 men and women scientists and doctors from universities, hospitals, and research institutions all across the country. They received inputs from many more individuals both pro- and antinuclear, and they have attempted to weigh and balance all these inputs. Full agreement has not been achievable even with this prestigious body, but their report airs clearly the differences that exist. It is entitled *The Effects on Populations of Exposure to Low Levels of Ionizing Radiation: 1980* and is objective and authoritative and highly recommended. Two other fundamental sources are massive reports (over 700 pages each) published by the United Nations Scientific Committee on the Effects of Atomic Radiation. The first is entitled *Sources and Effects of Ionizing Radiation*

(1977), and the more recent sequel is entitled *Ionizing Radiation: Sources and Biological Effects* (1982). Much of what follows is drawn from these documents in condensed and simplified form.

As these reports emphasize, there are really only two areas where the effects of low levels of radiation are of concern. First is the extent to which low levels of radiation increase the probability of occurrence of various forms of cancer. Second is whether low levels of radiation can cause changes in the genes controlling man's hereditary characteristics and thus result in birth defects or other harmful changes in our children or our children's children. We will look at this genetic question first.

RADIATION AND GENETICS. There is information to be gained on this topic from experiments on animals and from observation on direct human experience, and we will try to learn from both. First—the experiments on animals.

Experiments have been carried out on dogs, rabbits, rats, hamsters, and particularly on mice. Even man's nearest relatives the monkeys have been used to study radiation effects. These studies have tried to learn about the effects of radiation on the embryo and foetus before birth and the effects on sperm and egg before conception. We want to know what happens when a pregnant mother is accidentally subjected to unusual amounts of radiation, and we want to know whether a man or woman need worry about effects on children conceived at any time up to many years after an accidental irradiation. The first question thus concerns the direct effects of radiation on an embryo or foetus; the second question concerns the effects on the genes or chromosomes, the chemical units carrying the hereditary traits of the father or mother.

When a sperm and egg join, the cells begin to multiply to gradually form and shape the image that is dictated by the gene and chromosome pattern received from the 2 parents. Initially this growing embryo, begun by fertilization in the tube connecting the woman's ovaries to the uterus, is free to move down the tube to the uterus. There in the first few weeks it attaches to the lining of the uterus where it continues to grow until the marvelous moment of birth. This act of implantation or attachment to the uterus wall is a critical event, and radiation received after fertilization but before implantation can result in the death of the embryo or failure to implant, which has the same fatal result. Experiments on animals, however, have shown that if the embryo survives the radiation and becomes successfully implanted, its growth from that stage on will be normal, and there is no evidence of an increased number of birth defects or later problems.

If irradiation occurs after implantation and during the critical period of 9 to 40 days into the embryonic or foetal life, much more serious problems can arise. This is the period when the limbs and body organs are being formed in the marvelous pattern programed by the genes and chromosomes. Severe radiation doses in this period can disrupt the delicate process of synthesis and formation and can result in malformations or failure of bodily functions. Doses as low as 100 rads will cause death either in the uterus or at birth in half of the cases thus irradiated. In mice, doses as low as 5 rads can cause some cases of malformation but whether this level has similar effects in man is not clear. The data from Hiroshima and Nagasaki, which have been discussed in this chapter, showed that irradiation in the period from 3 to 17 weeks after conception showed clear evidence of increased frequency of mental retardation at levels of 50 rads and above. Various studies of women given radiation treatments in the same period of their pregnancies but at levels of only a few rads have not shown significant increases in malformations, however. Thus it is clear that in this early period of pregnancy the embryo or emerging foetus is particularly sensitive to damage from radiation. The damage seems to begin at levels between 5 and 50 rads and be greater as doses increase. The question of whether subtle effects exist at levels below 5 rads is unanswered; if the effects exist, they are small and relatively rare.

In later stages of pregnancy the foetus, now well formed, seems more resistant to harm from radiation but still more susceptible than an adult or an infant after birth. Studies of children who were irradiated in the foetal stage at Hiroshima and Nagasaki showed that doses above 50 rads resulted in clear evidence of smaller body size, for example, when examined when they were 17 years old.

Radiation damage to the human form before birth is thus well documented at high doses but not definitely established at doses below 5 rads. We would also like to know whether radiation received by a man or a woman will have permanent effects on the chromosomes and genes they produce and hence will change the inherited characteristics of children conceived *after* the time of irradiation. Can the effects thus be multiplied generation after generation? This is the heart of the question of whether radiation produces genetic damage.

In seeking an answer to this question we must first understand how inherited characteristics are passed on from parents to children through the intricate chemistry and structure of the chromosomes. Our body cells and the germ cells producing sperm and egg each contain a pair of chromosomes in turn containing at their core a long spiral thread of complex chemical material known as DNA. Segments of this long chain of DNA are the genes that carry the chemical coding for particular hereditary characteristics. In the formation of sperm or egg the chromosome pairs separate so that only one chromosome exists in sperm or egg.

In this separation there is a mixing of the genes so that there is a scrambled selection in the sperm cell, for example, including on a random basis half the genes from the male parent. A similar selection occurs in the egg from the female, and in the new union of sperm and egg this wondrous process results in a new pair of chromosomes in each cell with random selection of genetic codes, half from each parent.

There are many ways in which damage can occur to the chromosomes and genes of this complex process, resulting in undesirable changes in the offspring. There can be a change in the character of one of the genes themselves. This is known as a mutation, and it may result from chemical effects, radiation, or other causes. There can be rearrangement of parts of the chromosomes carrying many genes, or there can be too many or too few chromosomes. Down's syndrome, causing what was known in the past as as mongoloid child, is the result of a particular extra chromosome. Radiation does not seem to be a significant cause of errors in chromosome distribution but causes genetic change to some extent by inducing rearrangement within the chromosomes and particularly by causing gene mutations. The discussion that follows will focus on mutation effects but will include other effects as well in considering the overall impact of radiation.

Mutations are the cause of many common genetic problems including physical abnormalities, proclivity to certain diseases, or abnormal behavioral characteristics. The mutation can be passed on from generation to generation, thus perpetuating the defect, or it can die out because the change it induces is so severe the individual carrying it cannot survive or reproduce. Probably 1 out of every 10 children born has some significant adverse characteristic because of a genetic mutation. It has been estimated that possibly 1 to 6% of these mutations have originated because of exposure to natural background radiation at an average level of 100 mrems/yr or 3 rems/generation (assumed to be 30 years). This is a very difficult number to authenticate, however, since as the latest BEIR report states, "There has been no unequivocal demonstration of radiation-induced gene mutation in humans, and thus there are no data on induced mutation rate."

If that last statement is true, you may ask why there is so much concern about the genetic effects of radiation. It is because even though we have no absolute evidence of its existence in humans, we have such strong evidence of its existence in insects and animals that we know it must be there. The irradiation of generations of a fruit fly called Drosophila has shown definitely that high levels of radiation can induce mutation effects that are passed on from generation to generation. Probably the most convincing evidence has been that obtained from experiments on mice. In one of these tests the sperm-generating cells in mice were subjected to a complex dose pattern consisting of first 100 rads (at 60 rads/min), then a 24-hour wait and finally another dose of 500 rads (also at 60 rads/min). Of 2646 male offspring born to these parents as a result of mating after the irradiation, 37 were found to have skeletal defects that were passed on to succeeding generations. Typical defects were too few or too many bones, changes in bone positions, or fusion of bones. This is a mutation rate of 1.4%, but it must be remembered that these are very large doses; that the results are on mice, not humans; and that the split radiation dose was used because it was known to produce maximum probability of genetic changes.

There has been much more experimentation on mice and other small animals largely relying on microscopic evidence of genetic damage. The final conclusion of the BEIR Committee in 1980 was that considering all the evidence "the probable increase in incidence of dominant genetic disorders that lead to serious handicaps at some time in the life amounts to 5–45 per million liveborn" for 1 rem of irradiation to the sperm cells of male parents. Assuming a natural radiation level of 3 rems per generation, this would imply 15 to 135 such genetic disorders per million children being caused by that radiation. We said earlier that 1 out of every 10 children born has some significant adverse characteristic because of a genetic mutation. This would be 100,000 such defects per million children, so new mutations introduced by radiation are only a tiny fraction of the harmful mutations that exist. There is still the fact that new mutations from one generation are mostly passed on to the next generation, and the total number in existence originated by radiation continues to grow until an equilibrium number is reached. This equilibrium is reached because in each generation some of the harmful mutations are *not* passed on, and this bad strain disappears forever. Eventually the number disappearing becomes equal to the number of new ones generated, and the total number in existence stays constant as long as the basic natural radiation level remains constant. Geneticists now estimate that of all the harmful mutated genes now existing in man about 1 to 6% had their origin in effects of natural radiation. Ninety-four to 99% originated from other causes. Thus while radiation is a contributor to harmful genetic change, it is a minor one.

We are led to ask, however, how additional amounts of radiation generated by man will change this pattern. One way of understanding this is to determine how much radiation would be necessary to double the number of induced mutations and hence increase radiation's contribution to a range of 2 to 12% of all harmful mutations, still a small percentage. Largely from data from experiments on mice this has been calculated to be 50 to 250 rems, a very large dose. Data collected for many years on the children born of parents irradiated at Hiroshima and

Nagasaki show a slight indication of an increase in genetically induced defects. The effect is evident only in the few children from the small number of people receiving the highest dose levels, and even there it is too small an effect to be sure it is real. If we assume it *is* real, the dose to cause doubling of the effects caused by natural radiation would have to be 156 rems when received as a single short burst. This seems to confirm what would be expected from animal data, but it is not by any means an "unequivocal demonstration" of genetic radiation damage in humans.

Tests on animals show quite clearly that genetic damage can occur as a result of very large doses of radiation. They show also that doses below 1 rem probably, but not certainly, cause a very low level of genetic damage, an effect largely lost in the background of other causes of genetic change. Such data as exist from direct experience on man also confirm that if genetic effects exist from low levels of radiation, they are of far less consequence than other sources. This is counter to the public misconception that even small amounts of radiation cause serious genetic harm. They clearly do not. Perhaps in time as more valid experiments are carried out and reported the detailed facts will emerge more clearly. The basic truth that low levels of radiation cause little if any genetic damage in humans will not change, however.

RADIATION AND CANCER. The initiation of various forms of cancer is by far the most important harmful effect of radiation on the cells of the body. We pointed out earlier that cells may be killed, they may be damaged but heal, or they may be damaged in such a way as to undergo the abnormal growth and multiplication we lump under the term cancer. There are, of course, many different kinds of cancer, varying with the cell type affected and the part of the body in which it resides and functions. There is no way to trace a cancer to radiation as its source. The cancer cells will appear the same whether they were initiated by radiation, by toxic chemicals, by viruses, or by some other aberration of cell function.

When we begin to look at the details of how radiation results in cancer we soon recognize that certain tissues are more sensitive than others towards cancerous reaction to radiation. The thyroid, the lung, some of the digestive organs, and a woman's breasts are particularly susceptible, for example. The probability of cancer induction is highest when there is actual absorption of the radioactive source in the body. The ingestion of radioactive iodine, for example, leads to its concentration in the thyroid gland, and, if the amounts are high enough, can induce cancers there. Radioactive particles such as the disintegration products from radon can lodge in the lung and eventually cause lung cancer, just as cigarette smoke does from its chemical contamination.

If we are interested in fine details we must distinguish between the effects of the different types of radiation. Here we can introduce a new term linear energy transfer or LET. Particles, such as α particles, which lose a lot of energy to the tissue through which they pass in a short distance, are said to have high LET and a high likelihood of causing damage to the tissue. Neutrons are also high-LET particles, even though they have high-penetrating power. Beta particles, X rays, and γ rays deposit less energy in the tissues through which they pass and are thus low-LET radiation. You will recall that the rem is a unit designed to equalize these differences in rates of energy deposition, so that the effects of 1 rem of α particles are the same as 1 rem of γ rays. For low-LET radiation 1 rem is approximately equal to 1 rad, and in discussions of radiation effects at Hiroshima and Nagasaki we have used rem and rad interchangeably since almost all the biological damage was from γ radiation, which is low LET. In considering the effects of low levels of radiation these distinctions as to type of radiation, its source, and its rate of energy transfer may be important.

Another facet of radiation in its role as originator of cancer is the question of whether a single dose, 50 rems or 50 rads, for example, can carry the threat of cancer for many years afterward. How long is one in jeopardy? For leukemias there seems to be an induction period of a few years before first appearance of symptoms, and the probability of occurrence appears to peak perhaps 10 to 15 years after the irradiation. After 25 years the probability of leukemia developing seems to drop back to the natural rate that would have occurred in the absence of radiation. This peaking effect seems to be seen in the statistics from Hiroshima and Nagasaki, but as we have seen in the figures, the number of cases is so small that it is hard to be sure the effect is real. If one is talking radiation levels of 10 rems or below, there is just not enough evidence to tell one way or the other.

Leukemia is essentially a cancer of the blood stream and is to be contrasted with solid cancers, those which arise as tumors or lumps in a fixed location in the body. Solid cancers have long latent periods and seldom appear less than 10 years after the radiation exposure. Why? We really do not know. Similarly they may still appear as much as 30 or more years after the exposure; so, unlike leukemias where the threat diminishes after 20 to 25 years, the threat of solid cancers probably lasts a lifetime.

But how great is this threat when we are considering low levels of radiation? That is the critical question. It is an extraordinarily difficult question to answer for reasons we have discussed before but bear repeating. First and foremost it is because the effect is so small it cannot be measured directly. The evidence is confounded by the existence of so many other causes for cancer that are clearly more to blame than radiation. We have too little knowledge about how cancer begins to be able to establish clear models relating the rate of cancer causation at high-radiation levels, which we *can* measure, to rates of causation at low levels of radiation, which we *cannot*. These models are at the heart of the controversy over the effects of low levels of radiation, and we must now examine both the models and the controversy.

By referring to the data on the leukemia death rate of people irradiated at Hiroshima it is easy to see that above 50 rads the death rate increases with increasing radiation dose. At lower dose rates, however, the death rates were actually *less* than would have been expected in an unirradiated population of the same size. Does that mean that low levels of irradiation *prevent* leukemia? That seems very unlikely, but it probably means that the effect is so small that it is masked by other differences in life-style of the people, their original health level, or some other statistical variation. The number of people irradiated, even though numbering in the thousands, was too small a sample to detect the tiny effect that probably existed.

This means that the only tools we have for predicting the cancer-causing effects of low levels of radiation are models through which we extend the known effects of high levels of radiation to the probable or possible effects at low levels. When we look at the high-level effects we find that in some cases the increase with dose is linear; if the dose is doubled, the frequency of cancer cases is doubled, or if the dose is increased 4 times, the number of cases of cancer goes up 4 times. In other cases there is a greater sensitivity to dose, and a doubling of dose will cause 4 times as many cancers while 4 times the dose will cause 16 times as many cancers. This is called a quadratic response, where in mathematical terms the effect increases as the square of the dose. These responses are measurable by experiment, but how can we tell what is happening when the doses become smaller and it would take tests on millions of people or animals to get any meaningful statistics?

From the Hiroshima data, for example, we can detect an increase in the incidence of leukemia down to 50 rads, but below this the data are not meaningful. In the narrow range where it can be measured, the effect is nearly linear with dose. If we assume it *remains* linear to smaller and smaller doses, calculation tells us that 750 millirads would result in 1 death from leukemia in a period of 25 years in a population of 32,500 receiving that much radiation in a single dose. Japanese statistics tell us that in an unirradiated population of 32,500 people over 25 years there will be about 70 deaths from leukemia. Thus a 750 millirad dose *might* increase the leukemia death rate by 1.4%, but we can't be sure.

We could assume that the effect below 75 rads was not linear but quadratic and that at 1/10 the dose (i.e., at 7.5 rads) the effect would be only 1/100 the effect at 75 rads and at 750 millirads it would be only 1/10,000 the effect at 75 rads. In that case 750 millirads of radiation would induce only 1 death over the same period of years in a population of 325 million people. Why isn't this assumption just as valid as the linear one discussed above? Many reputable scientists believe that this model, the quadratic model, is in fact the one most likely to be true. Others argue that since we can't really tell, it is safer to assume that a linear response exists since this would predict a higher rate of cancer induction and would result in more stringent and more certain standards for allowed levels of radiation.

There are some scientists who will even argue that perhaps at low levels of radiation the sensitivity for cancer induction increases and if the radiation levels is reduced by a factor of 10, for example, perhaps the rate of cancer induction only drops by a factor of 5 or less. Their argument is based on the following theory. We know that at high levels of radiation we may get killing, cancer induction, or damage followed by healing. As the level of radiation drops it will become insufficient to kill cells but can still result in either cancer induction or complete healing. Since we don't know just what triggers cancerous growth we can't rule out the possibility that low-irradiation levels might have greater capability for this triggering action, and hence cancer induction could continue to be significant even down to very low levels.

The National Academy of Sciences BEIR study examined carefully all the evidence pointing to such a pattern of behavior. Their conclusion was that none of the evidence reliably supported such a thesis and that even a linear model, in the case of external radiation, was probably overconservative. Where internally absorbed high-LET radiation such as from internally deposited α-emitting particles is concerned, the linear model may be less conservative.

SOME AREAS OF CONTROVERSY

Concerns over the effects of low levels of radiation have led to a number of areas of bitter and often well-publicized controversy. To really understand the issues we must examine at least some of them directly to get them in perspective.

HANFORD WORKERS. One of the areas of controversy involves a series of studies made on employees of the Hanford Works at Richland, Washington. This site began operation during World War II as the location for the huge graphite-moderated nuclear reactors designed to produce plutonium for the U.S. weapons program. Operation of the reactors began in 1944, and the site also included plants for the chemical separation of the plutonium produced in the reactors and plants for the production of weapons parts. In later years plants were built on the same site for electric power production. There has been continuous activity at Hanford involving low-level occupational exposure to radiation for more than 40 years. It seemed logical, therefore, a number of years ago, that a study of the medical history of the people who had worked at Hanford might yield valuable information on whether or not there were identifiable increases in frequency of occurrence of cancer in workers who had received radiation doses significantly above background. The Atomic Energy Commission therefore awarded a contract in 1964 to Dr. Thomas Mancuso of the University of Pittsburgh to examine the data available and see what conclusions could be reached.

The study continued over more than a decade and resulted finally in the publication in 1977 of a very controversial paper in the magazine *Health Physics* under the authorship of Dr. Mancuso and Drs. Alice Stewart and George Kneale. Their procedure had been to obtain the death certificates of as many male workers as possible who had worked for an appreciable period at Hanford after 1943 and had died sometime between 1944 and 1972. Since very large numbers of workers had been involved, they found 3520 deaths to have occurred, a substantial sample with which to work. The records showed that 2184 deaths were of people who had received some radiation exposure during their period at Hanford and 1336 were of people who had received no radiation exposure. The question was whether cancer was a greater cause of death in the exposed group than it was in the unexposed group. Mancuso, Stewart, and Kneale believed the data showed a definite increase in cancer rate for the exposed group. To them the data indicated that for every rad of radiation received the general cancer rate would increase by 8%, while more specifically the rate for pancreatic cancer would increase 14%, lung cancer 16%, lymphatic and blood cancers 40%, and bone marrow cancers 125%.

These data indicated a much higher causal effect for cancer from low levels of radiation than had been evident from the studies of survivors of Hiroshima and Nagasaki, and there was immediate and strong reaction to the report. The reaction included the scheduling of congressional hearings by concerned legislators and a series of critiques and additional analyses by skeptical scientists. An analysis by G. B. Hutchison and others, for example, using the same data as analyzed by Mancuso and his associates, found there was no correlation between level of radiation exposure and total deaths from cancer. Their analysis gave careful consideration to factors such as length of exposure time, age at death, and other significant variables in the lives of the individuals studied. Other studies analyzed the mortality data on nearly 21,000 white males who had worked at Hanford before 1966. No correlation was found for lung cancer and total cancer deaths. Both these studies and the Hutchison study referred to above did show a statistically significant increase in cancer of the pancreas and in multiple myeloma (a cancer of the bone marrow). There were only 14 deaths from pancreatic cancer and 4 deaths from multiple myeloma, however, so the numbers were very small. The Academy of Sciences BEIR report concluded that even these numbers are probably not valid, based on the small number of cases and the absence of indications from other sources that these types of cancer are dramatically increased by exposure to radiation.

What then does all the analysis of data on Hanford workers tell us? First it points up the difficulty of obtaining unassailable conclusions from analyses of even fairly large populations exposed to low levels of radiation even fairly continuously over relatively long periods of time. Second it emphasizes again that if effects exist from radiation amounts below 10 rads, they are so small as to be swamped out by other differences in the environment in which the subjects live. The Hanford study was complicated, for example, by the significantly lower than normal overall death rate for the group of people studied. Because of the nature of the selection process by which they obtained their jobs, the climate in which they lived,and the health care they received, the Hanford workers were healthier than the national average. This fact, and its influence on the statistics, was overlooked in the first analysis but included in the later ones. Finally the Hanford studies tell us the essential need for dispassionate analysis considering all the facts. It is unfortunate that the dramatic nature of the early findings resulted in much publicity and amplified unnecessarily the concern over the effects of low levels of radiation. The Mancuso, Stewart, and Kneale findings are still used as gospel by many of the groups opposed to nuclear power without recognition of the tempering effect of later analyses.

PORTSMOUTH NAVY YARD STUDY. Another study that received much publicity and is still often quoted today is by T. Najarian and T. Colton on radiation effects on employees of the Portsmouth Naval Shipyard in New Hampshire. This study was based on interviews with near relatives of 525 people who had been workers at the shipyard and had subsequently died. The shipyard had been involved in building and servicing nuclear submarines, and the workers all wore monitoring badges from which records on radiation exposure had been compiled. Najarian and Colton interviewed the next of kin and determined from them whether the deceased had worked with radiation or not. Based on these interviews they determined that the number of leukemia deaths was more than 5 times normal (6 versus 1.1), all blood cancers were over 3 times normal (10 versus 2.9), and deaths from all cancers were twice the normal frequency (56 versus 31.5 expected). These findings created so much furor that the shipyard finally released employment and radiation statistics for all 1722 people who had died since their employment. This permitted a more careful and exact study to be made based on actual records rather than hearsay. The number of deaths from cancer generally was not related to the radiation dose received. Only for cancers involving the blood stream was there any indication of a trend with dose. For cumulative doses above 1 rem, cancers of the blood were the cause in 6 of the 113 deaths of individuals receiving this much radiation. Normally for 113 deaths at the same ages, 2.3 would have been from this type of cancer. Statistical analysis showed that with this small sample the result could not be called positive evidence. The National Academy BEIR report reached the following conclusion: "These analyses of proportional mortality among Portsmouth Naval Shipyard workers contribute little to our understanding of health risks from low-level radiation. However, they do provide a remarkable illustration of the dangers of response bias in epidemiological studies."

WEAPONS-TEST FALLOUT. Fallout from nuclear weapons testing has added significantly to the world's background of radiation. This has inevitably resulted in many studies to determine whether such fallout has actually caused illness or death, and equally inevitably, on the part of those directly in the path of such fallout, has resulted in numerous claims of just such damage. One of the primary sources of weapons-test fallout in a relatively populated area has been the U.S. weapons tests carried out at the Nevada Test Site in the desert near the Nevada-Utah boundary. Ninety-seven above ground tests were made at the site between 1951 and the end of 1958. Of these at least 26 resulted in wind-carried distribution of fallout in the state of Utah. Many years later Dr. Joseph Lyon and some of his associates at the University of Utah College of Medicine undertook a study to determine, if possible, whether any measurable consequences could be found. They reasoned that the individuals most susceptible to harm from this fallout would be the small children then living in the area. They also recognized from their knowledge of radiation effects that the most probable effect would be an increase in the number of leukemia cases in the area receiving the most fallout. Their study therefore focused on leukemia and other childhood cancers occurring in Utah between the years 1944 and 1975. Through the state registrar of vital statistics, death certificates were obtained for all children who had died younger than 15 years old during this period. This showed that during those years there had been 357 deaths from leukemia and 386 deaths from other childhood cancers. Now the question was whether an unusual number of these were residents during 1951 to 1958 in the high-fallout counties.

Maps were obtained from the U.S. Defense Nuclear Agency and from the Energy Research and Development Administration showing the probable fallout pattern from all of the tests. The maps were used to identify 17 counties as "high-fallout counties," and all other counties were lumped as "low-fallout counties." Based on this and times of residence the deaths were assigned to a "high-exposure cohort" or a "low-exposure cohort." This inevitably meant that the high-exposure cohort consisted almost entirely of children from rural areas while the low-exposure cohort included large numbers of children from larger population centers.

To understand the results we have to consider the records of three different groups of children:

GROUP 1: born in the period 1944 to 1950 before the nuclear tests and hence not exposed to fallout at their youngest age but possibly exposed, depending on where they lived, in the age range from 1 to 15 years old.

GROUP 2: born in the period 1951 to 1958 and hence possibly exposed from foetal age to 8 years old.

GROUP 3: born from 1959 on and hence not exposed to any new radioactive fallout from atmospheric tests.

In Group 1 there were 51 deaths from leukemia before the age of 15. Seven of these were in the counties that were the high-fallout counties and 44 were in counties designated as low-fallout counties. On a rate basis this was 2.10 fatalities per 100,000 people in the high-fallout counties and 3.84 fatalities per 100,000 people in the low-fallout counties. The national average for that period was about 3.6 per 100,000 for the same age group, so the rate in the high-fallout counties was well *below* the national average.

In Group 2 there were 184 leukemia deaths prior to age 15 of which 32 were in high-fallout counties and 152 in low-fallout counties. Rates were calculated as 4.39 per 100,000 in the high-fallout counties and 4.21 in the low-fallout counties. (The population level was very much larger in the low-fallout counties than in the high-fallout counties.)

In Group 3 (born after the weapons tests) there were 122 leukemia deaths, but on a rate basis there were only 1.96 deaths per 100,000 population in the high-fallout area and 3.28 in the low-fallout counties.

When the timing of the deaths was examined it was found that if 4 separate time periods were considered (1944 to 1950, 1951 to 1958, 1959 to 1967, 1968 to 1975), the death rate from leukemia in children in the high-fallout counties was below that in the rest of Utah and the United States for all but the 1959 to 1967 period when it reached a level of about 6 per 100,000 population as contrasted to a prior level of slightly over 2 and a national average of about 4.

These are complicated statistics, but they seem to be saying the following:

1. Those counties directly in the path of the weapons-test fallout, in the absence of fallout, had a naturally low rate of incidence of childhood leukemia.

2. The fallout in the years 1951 to 1958 caused an increase in the rate of induction of leukemia in very young children that showed up as an excess number of deaths in the period from 1959 to 1967. How many? Perhaps as many as 16 (32 actual versus 16 expected at the lower rate of earlier and later years).

There are many uncertainties in this picture, however. Could there have been some other cause? Why weren't there increased deaths in the group born prior to the testing but still at very young ages when the tests were under way? The numbers are still pretty small and we can't be sure they are real. Probably the greatest uncertainty is that we have no way of knowing what the exact dose levels were or whether some few children might have received much larger doses than others. Gamma doses totaling several rem or more seem quite possible, with some of it from ingestion into lung or digestive system.

These data tend to confirm the facts learned from Hiroshima and Nagasaki that radiation can cause an increase in the rate of occurrence of leukemia, particularly when exposure is at a very early age. They do not help us much, however, in determining whether this risk is still present to some degree at very low radiation levels or whether there is a level below which there is no measurable harm.

OTHER EVIDENCE. There have been numerous other studies attempting to determine the extent to which low levels of radiation will affect the probability of occurrence of cancer. In one such study by Dr. Alice Stewart and associates the records were obtained for all children in England, Scotland, and Wales who had died of cancer between birth and 9 years of age during the period 1953 to 1965. Slightly over 7600 such cases were authenticated, and a study was made of the diagnostic X-ray exposure received by mothers of these children compared to a matching sample of a similar number of children who had remained healthy. Of those that had died, 1141 out of 7649 (14.9%) had received one or more obstetric X rays, while of the randomly selected live group only 774 out of 7649 (10.1%) had received such X rays. The conclusion drawn was that the obstetric X rays had increased the probability of death from early cancer by 47% (14.9 − 10.1/10.1 × 100 = 47.5%). Are these conclusions valid? We cannot be sure. Some critics have argued there may be other contributing factors that led to the obstetrician requesting X rays for the group that eventually died. Was the control group a valid match? Dr. Stewart's more detailed analysis of the data led her to conclude that even at these low-dose levels (200 mrads and up) the probability of cancer increased with dose, that for prenatal irradiation the added risk was the same for solid tumors as for leukemia, that 1 million exposures to 1 rad of radiation before birth would cause 600 cases of cancer, and that exposure early in pregnancy was more dangerous than later in pregnancy. We should not accept these conclusions as fully proven, but they are a valid part of the complex data in a complex area.

There are certain regions of the world where background radiation is substantially higher than normal because of the presence of surface rocks and soil containing thorium or uranium or both. One such area is in the Guangdon Province in China where the presence of a thorium ore called monazite causes a γ radiation background about 3 times normal. In the early 1970s a massive study was undertaken to determine whether there were any significant health or genetic effects in two of the highest radiation areas compared to a nearby area where radiation background was normal. Data were collected on the health and family history of 73,000 people in the high-background areas, some of whose families had lived there up to 16 generations, and about 77,000 people in the normal background area. About 20,000 people of each group were examined in some detail, and much more extensive evaluations were made on smaller groups of about 3500 people. Radiation levels were confirmed by having about 400 randomly selected people in each area wear recording radiation dosimeters for a period of 2 months. They wore them around their waists during the day and placed them on their beds at night.

On the average the people in the high-dose regions received about 231 millirem per year whole-body radiation and 437 mrem/yr on the tissues lining the bone cavities, while those in the low-dose region received an average of 96 mrem whole-body and 153 mrem in the bone cavity lining. The high values in the bone tissue resulted mostly from radium-228 and radium-226 taken into the body through water and the locally produced meat and vegetables. The dose values were calculated from actual analytical measurements made on the food and water. One can say roughly, therefore, that the dose received in the high-background region was 2.5 times as high as that received in the low-background region and that in both regions the dose was both external and internal.

Careful examinations were made for (1) chromosome damage that might indicate genetic impairment, (2) rates of hereditary disease and birth defects, (3) foetal deaths resulting in spontaneous abortion, and (4) frequency of various forms of cancer. There were no significant differences in any of these areas between the 2 groups, even in the area of cancer frequency. One odd anomaly appeared, however, in the statistics on hereditary diseases and birth defects. The overall rate of occurrence of 31 kinds of birth defects and hereditary diseases was lower in the high-irradiation group (13.7/1000 births) than in the control group (14.5/1000 births). There were 6 cases of Down's syndrome in the high-irradiation group of 3504 cases studied and none in the 3170 cases in the low-irradiation group. This might be a random variance because of the small number of cases involved, it might result from some other causative element, or it might have some relation to the radiation levels. Other studies should look for further evidence to substantiate or deny the possible effect. With this one possible exception the Chinese study tells us that increasing background-radiation levels by a factor of 2.5 does not cause any discernible health effects. But can we tell what 5 times or 10 times or 100 times will do?

Linear extrapolation of the Hiroshima data showed us that the high-level radiation results would predict 1 extra case of leukemia in a group of 32,500 people receiving 750 mrem radiation in a single dose. If the effect is truly linear and remains so to very low levels of radiation, then one-tenth of this dose (75 mrem) should cause 1 case of leukemia over 25 years in a population of 325,000 people. We said previously that the conclusion of the Chinese study was that there was no significant difference in the cancer rates between the two groups. This was correct. Some types showed greater frequency in one group and some in the other as would be expected from random chance. The rate for leukemia, however, was 5.3/100,000 people in the high-radiation group and 4.1/100,000 in the low-radiation group. Is this real? We cannot tell, but considering the average annual dose increment at the bone marrow to be almost 300 mrem $(437 - 153 = 284)$, this would not be inconsistent with the Hiroshima prediction.

On the other hand a carefully controlled study at the famous Mayo clinic in Minnesota failed to find any relationship between levels of diagnostic X rays received up to a total of 300 rads (300,000 mrad or mrem) and the onset of leukemia over a 20-year period from 1955 to 1974. Over this period 138 cases of leukemia were noted, and the radiation history of these individuals was matched to that of 276 controls (i.e., individuals of similar age, sex, and residential area who came to the clinic at about the same time as the patient in whom the leukemia was diagnosed). Since these patients had essentially all of their medical care provided by Mayo, accurate records existed of all the diagnostic radiation they had received. If these diagnostic X rays had been instrumental in causing the leukemias, the leukemia patients would have been expected to have received more X rays, on the average, than the controls who had not contracted leukemia. This was not the case, and the conclusion of the study was that diagnostic radiation in small amounts but up to as much as 300,000 mrem accumulated over a period of years did not increase the chances of an individual contracting leukemia.

We are thus faced with conflicting evidence but some clear facts. The evidence from Hiroshima and Nagasaki shows without question that a large single dose of predominantly γ radiation above 50 rem will increase the likelihood of leukemia and probably other cancers. Damaging effects to foetuses of women in early stages of pregnancy begin to appear at between 5 and 50 rems of X-ray exposure. At levels of exposure below 50 rems in adults the cancer-producing effects become difficult to quantify. It is clear that the lower the dose the smaller the effect, but the exact relationship between dose and level of effect is impossible to quantify precisely. There seems to be no valid evidence that low levels of radiation are more damaging per unit of dose than are high levels, and there is some evidence that shows the damage per unit of dose may be less. The conservative judgement, and the one used in setting standards for tolerable amounts of radiation, is that the damage per unit of dose is the same at all levels (i.e., that a linear relationship exists between dose and frequency of cancer induction). It is probable, but not certain, that single large doses are more damaging than the same amount received in many small doses accumulated over extended periods of time. It must be emphasized again that when we are talking radiation levels of a few rems (a few thousand mrem) or below, the effects are small and we are considering the probability of a few cancers in tens or hundreds of thousands of people. These are not high risks in our risk-filled world. They should not be ignored, but neither should they be exaggerated out of realistic proportion.

BOTTOM LINE: LOW LEVELS OF IONIZING RADIATION

All the evidence on ionizing radiation from natural and artificial sources points to only two identifiable effects from low levels of ionizing radiation—genetic effects and the initiation of cancer. The genetic effects on humans, if they exist at all, are very small and are probably measurably increased only by very large doses (50 rems or above). In our concerns about low levels of ionizing radiation, genetic effects should not be prominent. There is, however, a definite linkage between low levels of radiation and the induction of cancer. The effect is small and is of lesser consequence than that from many other causes such as carcinogenic chemicals, variations in body chemistry, possible viruses, and other as yet unidentified initiators.

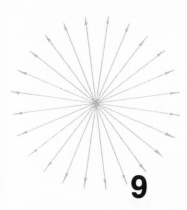

9

NONIONIZING RADIATION

In the Atom's Eye View we indicated that the only clearly proven damage effect of nonionizing radiation results from the higher internal body temperatures that can be generated. We need to enlarge upon this and see how much radiation is needed to cause harmful temperature rises and what the physical effects of these rises are in the humans in which they occur. We can then consider also the possibilities of more subtle effects at lower radiation intensities.

EFFECT OF MICROWAVES

Microwaves are particularly significant from the health standpoint because they are in a frequency range that penetrates and is absorbed in the human body. At either end of the microwave region the radiation is less damaging. At shorter wavelengths in the infrared region the radiation is absorbed only in a very shallow skin depth, and excessive radiation will result only in skin burns. Sunburn is a typical example. At the other end of the microwave spectrum, beyond 30 meters in wavelength, the radio waves are reflected without any penetration or absorption. Within the microwave region itself, between 100 gigaherz (GHz) and 10 megaherz (MHz), the extent of penetration and absorption varies with frequency and is maximum a little below 1 GHz. You will recall that UHF television lies between 432 and 728 MHz and is thus close to the region of maximum absorption. Microwave ovens are either 915 MHz or 2450 MHz and are also close to the region of highest absorption. Their purpose is to maximize absorption of microwaves to carry out their heating function. But the efficiency of energy absorption varies with the size of the body in which the radiation is being absorbed, so with roasts or steaks to be cooked the best efficiency is at somewhat higher frequencies than with larger objects such as the human body. There is a "resonant frequency" for maximum energy absorption that depends on size as well as the nature of the material in which the absorption takes place. Thus as one investigator put it, the "maximum deep heating for a fruit fly occurs over 10 GHz—hamburger at about 2.45 GHz—turkey at about 915 MHz and man at about 100 MHz."

The human body maintains an even temperature at any point by balancing the rate of heat generation from the utilization of food (metabolism) with the rate of heat loss through the tissue by conduction and the rate of heat removal by the blood stream flowing through the tissues. Eventually this heat is lost outside the body through the air we breathe in and out of the lungs and through the heat loss to the outside air through the skin. We have quite a wide range of control, particularly through expansion or contraction of the tiny blood channels (the capillaries) at the skin surface. The panting of a dog on a hot day is another way of getting more rapid heat removal. There are limits to the extent of this control, however, and if we have an unnatural source of internal heat generation, such as microwaves, we can exceed the rate at which heat can be removed. When that happens the body temperature rises above the 98.6°F, which is our normal balance point.

When you are sick you may have a temperature of as much as 103° or 104° without extreme concern on the part of your doctor. Higher than this, however, you can be in serious trouble. The addition of thermal energy (heat) within the body by microwaves in sufficient quantity can exceed the body's ability to cool itself and result in temperature increases well beyond 105°F. Tests in rats have shown that when temperatures reach 109 to 110°F, convulsions and death can follow if the radiation is continued. This is the extreme reaction from excessive amounts of microwave radiation, and in mice it can be induced at power densities as low as 5 milliwatts per square centimeter (5 mW/cm^2).

This is probably as good a time as any to digress into a discussion of measurement units again, for here as in many other aspects of radiation they can be confusing. We have defined the power density of a microwave beam as the amount of power passing through a 1 square centimeter area of space. Typically we speak of watts per square centimeter (W/cm^2) or milliwatts per square centimeter (mW/cm^2) or microwatts per square centimeter (μW/cm^2). A watt is a power unit, however. It measures an amount of energy per unit of time and is defined as a joule per second, where a joule is the international unit for a quantity of energy. Where we are concerned with the biological effects of microwaves we are really interested in the amount of energy actually deposited in that body, the number of joules per gram or kilogram, for example, or the rate at which that energy is being deposited, which is measured in watts per gram or kilogram. You may see reference, for example, to the Specific Absorption Rate (SAR) for microwaves, which is usually expressed as milliwatts per gram (mW/g) and refers to the actual rate at which the microwave energy is being absorbed in the body. This is the rate that really determines the amount of heating that occurs and the amount of harm done, if any.

The conversion from the power density of a microwave beam (in mW/cm², for example) to a SAR in mW/g is a complex one, because it depends on the part of the body involved, the orientation of the beam, the frequency of the beam, and other factors. It is thus difficult to say just what the physical effect of exposure to a given power density of microwave radiation will be unless all these conditions are known exactly. It is possible, however, to determine from the results of many experiments what the most serious effects of exposure to a given power density might be if all conditions tended to maximize the effect. When this is done it can be seen that the lowest levels at which convulsions can be induced in mice are about 5 mW/cm², in rats probably over 100 mW/cm², and in man considerably above that.

We have seen that body temperatures of 109°F or above are required to induce convulsions and possible death in mammals. It is also useful to see what levels of irradiation are just enough to increase the body temperature 1° or so, and whether there are harmful effects associated with these small temperature increases. In one experiment some small monkeys were irradiated with 2450 MHz microwaves to see how the radiation affected their ability to perform the task of pulling two separate levers to get pellets of food. Their body temperatures were carefully measured during the experiments. When the power density was 50 mW/cm² for 30 minutes a 1°C (1.8°F) rise in body temperature was noted but no loss in dexterity or activity in pulling the levers. When the power was increased above 60 mW/cm², however, performance with the levers began to be affected, and by 70 to 75 mW/cm² it was strongly disrupted. When the radiation was turned off, however, activity with the levers returned to normal. This tells us that with these monkeys measurable physical effects (temperature rise) occur at 50 mW/cm² and behavioral effects occur slightly above this level but appear to be temporary. Similar effects in man would be expected to occur at higher levels of power density because of his larger size.

In further confirmation of 50 mW/cm² as a safe threshold level for man it was found that rats irradiated with 2450 MHz microwaves at power densities from 35 to 65 mW/cm² for 30 minutes showed somewhat lowered endurance levels in a long-term swimming test begun immediately after the irradiation. Even with a 24-hour delay between the irradiation and the endurance test some loss of endurance persisted, but it was much less than in the test immediately after irradiation. Unfortunately tests were not continued to see if complete recovery occurred over a longer period of time, but from other evidence it seems probable that it would. Since rats are much smaller than man these tests could be construed as supporting 50 mW/cm² as a relatively safe level.

In the early history of high-power microwaves the main use was for radar for military purposes. The navy and the air force had the earliest serious concern with possible medical effects from microwaves, since their personnel were frequently involved where exposure was possible. The earliest basic standard was set in 1953 using some relatively simple assumptions. First, a calculation was made determining the average heat loss from a human body was 0.005 W/cm². It was assumed that when necessary the body is capable of dissipating very much larger amounts of heat than this, so an *input* of only twice as much as normal *output* should be safe. This led to a guideline that a person should not be exposed to microwave power densities greater than 0.010 W/cm², which is the same as 10 mW/cm². During the 1950s and early 1960s additional studies were made, but no evidence was uncovered warranting changing the guideline. In 1966 the American Standards Association (now called the U.S. Standards Institute) was asked to help draw up an official standard, and in that year issued a standard (C95.1) formalizing 10 mW/cm² as a safe limit.

There was not unanimous agreement as to the safety of this limit. Although no specific evidence could be cited for real effects in humans at levels this low, there were some cases of blood, testicle, or brain effects in rats at levels from 12 to 60 mW/cm². Some scientists felt these were too close to the allowed level for comfort, even though the results were on much smaller animals than man. There was also concern that there was not sufficient information on pulsed microwaves, as most radar was, compared to continuous wave radiation; and there was concern that too little information existed, even on animals, where low levels of radiation had been continued for long times.

Perhaps the most serious issue was whether effects from microwaves existed that did not result from higher temperatures in the body, but resulted from more subtle—possibly electrically related—effects on the eyes, the brain, or the nervous system. We know that the eye is a uniquely sensitive organ in our body, and this extraordinary sensitivity provides the incredible range of perception of light and dark, form and color that is the miracle of vision. Like many sensitive instruments, however, the eye is susceptible to damage; the very radiation it is designed to sense can be one of the damaging elements. Too much light can blind, and we have mentioned the serious problems of cataract blindness too often encountered by the early experimenters who lined up X-ray beams by eye and thus received large concentrated doses. A cataract is a "lessening of the transparency of the crystalline lens of the eye." It can come about because of changes in body chemistry that occur with age, it can be associated with diabetes or other diseases, or it can be induced by X radiation as mentioned above. Microwave radiation can also cause cataracts. There have been several well-documented cases of microwave workers developing them after exposure to relatively high levels of microwaves (in the order of 100 mW/cm^2). These levels and above have also clearly caused cataracts in the eyes of rabbits exposed experimentally. Whether lower levels for long times also cause cataracts is less clear, however.

One of the controversial figures in this issue is Dr. Milton Zaret, a New York ophthalmologist who was asked by the air force to examine the eyes of close to 1600 workers who had been associated with radar in the armed services. This survey was completed between 1960 and 1963 and led to the conclusion that there were no significant differences between the group exposed to low levels of microwave radiation and a selected control group who had not been exposed. Zaret later examined a number of patients who had cataracts and had records of previous exposure to microwaves. He became convinced that a particular type of cataract, one forming on the back of the eye rather than on the front lens, was being caused by microwave radiation and possibly by relatively low levels. It was in most cases impossible to determine actual exposure levels and times, and Zaret's views, lacking solid documentation, brought him into conflict with the same armed services who had earlier supported him. Perhaps more significant is that Zaret's findings were not substantiated by other investigators. Paul Brodeur in his somewhat inflammatory book *The Zapping of America* published in 1977 ascribes the lack of substantiation to a cover-up by the armed services in defense of their commitment to a 10 mW/cm^2 permissible standard. This appears to be an overdramatization of an admittedly defensive reaction by the armed services. The continued and open attempts since then to develop scientifically rigorous evidence of all sorts of microwave effects appears to refute the concept of a malicious cover-up. The slowness to revise standards in the face of growing evidence for such a need does at least confirm the existence of an undesirable degree of establishment entrenchment. As far as cataracts are concerned, however, the evidence still seems to show that damaging effects truly exist from microwave intensities of 100 mW/cm^2 and above and danger *may* exist from long-term exposure to levels as low as 5 to 10 mW/cm^2.

We now turn to the even more difficult question of microwave effects on brain function and the nervous system. It is a curious and not yet fully explained fact that microwave standards in the USSR and the Eastern European countries associated with it were set 1000 times lower than in the United States and Western Europe. Since the technologies utilizing microwaves (essentially radar) had grown up in military settings there was little if any communication between the USSR standard setters and those in the United States. The belief persisted in the Soviet Union that subtle nerve effects could exist that would make microwaves damaging to health at much lower levels than was believed to be the case in the United States. It is probably not possible today to give an absolute judgement as to where the truth lies, but it will be useful at least to see what the Soviet view is. It will also lead us to the subsequent discussion of the fascinating case of the microwave irradiation of the U.S. Embassy in Moscow.

In the early 1970s discussions were begun between the U.S. Department of Health, Education, and Welfare and the Ministry of Health of the USSR to "develop mutual cooperation in the fields of health and medical science." These led among other things to a joint cooperative effort to study the biological effects of microwave radiation. The program began in 1976 with the objective "to determine the nature, extent and main patterns of biological effects of microwaves as the basis for setting hygienic standards for levels in the environment." It was preceded by a mutual exchange of background information and an agreement as to the areas to be examined in more detail or to be confirmed by additional experiments.

As we know, the USSR and United States work prior to 1970 had been carried out essentially independently. The work in the United States had concentrated almost entirely on effects related to actual body temperature increases induced by short exposure to relatively high doses of microwaves. The Soviets, on the other hand, had explored a much more subtle range of effects they claimed could be observed after continuous exposure for times up to several months to microwave levels as low as 10 μW/cm^2, a factor of 1000 times less than the amounts deemed harmless by U.S. experimenters. The question was whether these differing views could be resolved by mutual discussion and experimentation.

While no complete consensus has yet been reached, it is clear that U.S. experimentation has confirmed at least part of the Soviet thesis that effects of microwaves exist that cannot be explained solely on the basis of the thermal effects. All of the data, both Soviet and United States, are from experiments on animals or animal tissue. There is no direct evidence of effects in man, and the effects in animals are subtle. We must at least recognize the existence of the effects and consider them in the establishment of safety standards.

The primary Soviet thesis is that long-term exposure to relatively low levels of microwaves can have subtle effects on the chemistry and electronics controlling the functions of our brain and nervous system. Their studies have focused on brain and nerve reactions and on behavioral characteristics of animals subjected to intensive microwave exposure. Typical of this approach was a study of brain-wave patterns in rabbits exposed to various intensities of 2375 MHz microwaves (about the frequency used in many microwave ovens) for a period of several months. Definite changes in patterns were observed at intensities as low as 50 μW/cm² with some evidence of effects at an even lower 10 μW/cm². This is a factor of 1000 less than exposure levels considered safe by U.S. standards, but there is no clear evidence that the pattern changes were harmful or that they persisted after the irradiation ceased. In another series of experiments using frog sciatic nerves (the main nerve in the leg) the reaction time of the nerves with and without microwave radiation (2.4 and 3 GHz) was measured. Small changes (10 to 15%) were observed at radiation levels near 10 mW/cm². This is the threshold for actual heating, but the experimenters found that heat alone could not cause the effects and thus assumed the microwaves were solely responsible. These experiments were duplicated in the United States with the conclusion that an effect from microwaves could be seen at 10 mW/cm² but not at 5 mW/cm². The damage at 10 mW/cm² appeared permanent, however, and the nerves did not recover full function when tested later in the absence of microwaves. It should be remembered that these tests were on nerves removed from the frog and irradiated while exposed in a special salt water solution.

Some of the most interesting and complicated experiments involved observation of behavior patterns in albino rats. A number of different tests were devised to measure the animal's reaction times and reflex responses. In one test, for example, an electric current could be passed through the wire bars at the bottom of the rat's cage. As the current was increased the tingling sensation in the rat's paws would make him begin to lift up his feet to avoid it. The threshold current at which this occurred could be used as a measure of the rat's alertness or sensitivity. In another more complicated test a bell and light signal were used as a warning that an uncomfortable pulse would be applied to the bottom of the cage 3 seconds later. A normal rat would eventually recognize that the bell and light meant that a shock was coming and that it could be avoided if he scampered to another section of the series of interconnected boxes in which he lived. The time necessary to learn this "conditioned reflex" action was used as a measure of the rat's mental sharpness. Duplicate setups were established with the only difference being that one of the setups included provisions for microwave irradiation of the rats at a frequency of 2375 MHz (2.4 GHz) for 7 hours a day over a period of 1 month for the highest intensity used (500 μW/cm²) and 3 months for the lower intensities (50, 10, 5, and 1 μW/cm²). Note that all these intensities are below the U.S. permissible levels.

Results of these and other experiments led the Soviet experimenters to conclude that "long term regular exposure of animals with EFD (energy flux density) of 500, 50 and perhaps 10 μW/cm² leads to more or less marked disturbances of various forms of inborn behavior and conditioned reflexes." A detailed examination of the data raises certain doubts about the firmness of these conclusions. The changes in response times are not very large (maximum 20 to 40%), for example, and for the test on sensitivity of detection of electric current the sensitivity is initially *increased* by intensities of 500 and 5 μW/cm² but is *decreased* by 50 and 10 μW/cm². The changes are thus not consistent in direction. Similarly in 1 set of the conditioned reflex experiments there appears to be no effect in the tests at 500 μW/cm² for 30 days but significant changes at 50 μW/cm² after 30, 60, and particularly 90 days exposure. Why should the higher intensity yield negative results and the lower intensity positive ones?

There is evidence of other types of effects as well. Another series of Soviet experiments studied the effects of microwave radiation on the blood chemistry of rats related to their immune reactions. This is the chemical response to bacteria or other harmful cells that may have invaded the body. The presence of these foreign intruders stimulates the development of special cells to fight the invaders, and the degree of development of these "fighter" cells can be followed by examining blood samples. Albino rats were again used as the subject of study. Irradiation levels of 500, 50, 10, 5, and 1 μW/cm² were used and results were compared with a control group receiving no radiation. Irradiation for 7 hr/day with 2.4 GHz microwaves at 500 μW/cm² for 30 days resulted in substantially lower numbers of fighter cells (called blasts), and decreases were seen as early as 3 days into the irradiation period. While recovery occurred slowly after the irradiation ceased, full recovery had not occurred even 3 months after the irradiation. Similar effects, although to a lesser degree, occurred at radiation levels of 50 μW/cm² and even 10 μW/cm², although essentially complete recovery seemed to take place within 60 to 90 days after the irradiation ceased. These changes appear to indicate that the rats' resistance to infection and disease is lowered by continuous radiation with 2.4 GHz microwaves for many days even if the levels are as low as 50 to 100 μW/cm². The data are for rats, not humans, however, and the true relevance to human disease resistance is far from clear.

In view of the potential importance of the Soviet findings experiments have been carried out in the United States at the University of Washington School of Medicine and elsewhere to try to verify the Soviet work. Carefully controlled experiments with rats exposed to 2.4 GHz radiation for as long as 3 months showed some, but not all, of the effects reported from the USSR. There were definite changes in blood chemistry related to levels of sodium, potassium, and carbon dioxide, but the difference disappeared 1 month after irradiation stopped. Behavioral differences were also observed similar to those described in the Soviet literature, but also with recovery in the absence of further irradiation. Another set of experiments, however, at higher intensity levels (2.5 to 10 mW/cm²) but at lower frequency (918 MHz) showed no evidence of blood chemistry or behavioral changes in adult rats. These studies led to the fascinating observation that permanent behavioral changes could apparently be induced in the offspring of female rats irradiated with microwaves during their pregnancy. The experiments that showed this were as follows. Eight pregnant female rats were irradiated for 20 hr/day with 918 MHz microwaves at an intensity of 5 mW/cm². Note that these are milliwatts and not microwatts and that this was a relatively high-intensity level. The offspring of the irradiated mothers were then compared with the offspring of 16 other control mothers kept in similar environments but without the irradiation. Physically all the young rats appeared essentially the same. At 90 days old, however, tests were begun to evaluate their ability to respond to a conditioned reflex test similar to the one used by the Soviets, which we described previously. A warning sound preceded an electric shock, and the percent of the time an avoidance maneuver (e.g. scuttling to another portion of the cage before the shock came) took place was used as a measure of the rat's mental dexterity. The offspring of the irradiated mothers, on the whole, responded only half as often as the offspring of the mothers who were not irradiated. The experimenters have suggested that this performance deficit could be related to lower levels of production of certain critical hormones from the glandular system of the offspring of the irradiated mothers. This is a tentative hypothesis the authors admit needs much more testing.

Taken together the USSR and U.S. data seem to indicate that there are in fact subtle effects of microwave radiation on nerve performance and blood chemistry and that these occur at levels possibly as low at 50 μW/cm² if the irradiation occurs over long periods of time. The groups responsible for setting acceptable limits for exposure to the average member of the public or the individual working with microwaves are wrestling with the significance of the data and the necessary changes in standards the data may imply. This is not an easy task because the uncertainties are great. The experimental evidence that exists is on tissue samples or on small animals. The data are frequently less than crystal clear, and one piece of evidence is often contradicted by another. The significance to a human being in terms of risk to health or well-being is very difficult to assess.

It is clear that current U.S. standards limiting permissible microwave exposure intensities to 10 mW/cm² are less conservative than was originally believed. It is also clear that such harmful effects as exist below that level are relatively subtle and certainly not life threatening. That they exist at all, however, leads us to several precautionary actions. First, we must review our present standards and make such realistic revisions as a rational balance beween risk and economics dictates. Second, we should use intelligent caution to minimize exposure to excessive amounts of microwave radiation, particularly in the frequency range from 10 MHz to 100 GHz. This doesn't mean giving up microwave ovens or TV; it just means prudence in design and use. Finally we should increase our efforts to understand more fully the real effects of microwaves on nerve function and body chemistry over wide ranges of frequency and exposure levels.

We can perhaps put this issue into better perspective by going back to Chapter 7 on sources of microwaves and relating commonly existing levels to the effects we have just been describing. You will recall that microwave power densities from 50 to slightly over 100 μW/cm² have been measured at One Biscayne Tower in Miami and 30 μW/cm² was measured at the observation windows on the 102nd floor of the Empire State Building. These levels are in the range where small effects were beginning to be observed in the experiments on rats just described. We do not know of any effects on humans at these levels, but we cannot unequivocally rule them out. Perhaps the most extraordinary attempt to determine whether such effects exist in humans occurred as a result of the bizarre story of the irradiation of the U.S. Embassy in Moscow. It is such a strange story that it is worth the detailed recounting that follows.

MOSCOW MYSTERY. The story begins in 1953 when personnel at the Embassy first detected the presence of a low-level microwave signal entering the Embassy from the outside. There appears to have been little concern about this radiation until the early 1960s when equipment was brought in to measure and record the arriving signals. The early lack of concern probably stems from the low signal strength and the absence of evidence from U.S. data that such low-exposure levels could be harmful. As the puzzlement about the reason for the irradiation grew, and the scientists began to hear more about Soviet beliefs that relatively low levels of microwave exposure could cause harmful effects, the level of concern also grew. In 1965, therefore, the Medical Services Office of the State Department conducted a survey of the medical records of 139 employees assigned to the Moscow Embassy and their 268 dependents. The study found no significant increase in abnormalities or illness but admitted that the nature of the study, based on less than complete records and without an unirradiated control group, made the results inconclusive.

The 1965 study was followed by a more detailed study undertaken by the George Washington University School of Medicine between March of 1966 and June 1969. It was called the "Moscow Viral Study" because most of the employees had still not been told of the presence of the radiation and a camouflage was needed. The main purpose of the study was to see if there was any evidence of chromosome damage indicating the possibility of harmful genetic effects in the subsequent children of the irradiated workers. Blood samples were taken at various times from more than 71 employees—some before their service at the Embassy, some during service, and some afterwards. The samples taken before service were designed to be the control set against which to compare results during and after irradiation. Again the results were inconclusive. Some evidence of chromosome damage was seen, or so the director of the study stated, in 43% of the samples during and after irradiation but were also seen in 38% of the control samples. These results were not thought to be statistically significant. In addition, an independent review of the data by three experts, one recommended by the principal investigator of the study and two by the State Department Medical Services Office, concluded that the blood sample microscopic slides were poorly prepared, difficult to interpret, and hence the results were highly questionable. Again, no clear answers. No proof of damage, but no proof there was none either.

In the early phases of the Moscow Viral Study it was recognized that it would be difficult to get valid data from a statistical study of the people at the Embassy. The irradiation doses would not be known with any certainty, and a matching sample of people who had not received radiation but were otherwise exactly similar in initial health characteristics and exposure to other environmental factors would be difficult to find. It was also recognized that experimental radiation on people was not reasonable. Monkeys, however, have many characteristics in common with people, and experimental radiation on monkeys was clearly possible. From this Project Pandora was born in 1965. The Advanced Research Projects Agency of the Defense Department funded the Walter Reed Army Institute of Research to carry out the work. Exposure intensities of 4 to 5 mW/cm² were used at frequencies matching those measured at the Embassy. Maximum levels at the Embassy had been only 18 μW/cm^2, a factor of 250 times lower. There were many experimental difficulties, and in 1968 a panel of outside scientists called the Science Advisory Committee was formed to review the data and see what they meant. The most important finding of the investigating team was that 2 monkeys showed "degradation in previously learned work performance" after exposure at intensities of 1 to 4.6 mW/cm^2 for 11 to 21 days for 10 hr/day. Subsequent experiments did not confirm this, however, and the Science Advisory Committee felt that somehow more specific information on humans must be developed. You might be able to tell something from a monkey's performance of tasks after irradiation, but you couldn't ask him how he felt.

The scruples about irradiation tests on humans were thus at least temporarily set aside and project "Big Boy" was commenced. It was not exactly a controlled experimental irradiation of humans, but it took advantage of the fact that certain crewmen who worked largely "above decks" on the aircraft carrier USS *Saratoga* were regularly exposed to microwaves from the ship's radar. Other crewmen who worked mostly "below decks" had little of this exposure but otherwise lived in a similar environment. Three sample groups of crewmen were selected: one receiving maximum levels of exposure (later shown to be from 0.3 to 1 mW/cm^2), one receiving generally lower levels of exposure, and one receiving no known exposure. The three groups were examined and tested for task performance, physiological effects, and biological effects. No significant differences were found between the 3 groups.

This study also showed that if effects did exist, they could only be proven by very careful studies using exactly measured radiation doses on human subjects under controlled conditions. The Science Advisory Committee gave serious thought to exactly how such studies could be organized and carried out. Ethical concerns eventually won out, and the final step to such experiments was never taken.

Two more studies must be reported to make the record complete. There has been evidence in Moscow Embassy employees of a higher than normal count of a type of white blood cells called lymphocytes. These, as mentioned previously, are the infection fighters, and high counts can indicate either the presence of some invaders such as viruses, bacteria, fungi, etc., or can indicate the presence of leukemia. For this reason a study was made between February 1976 and June 1978 comparing the lymphocyte count of personnel at the Moscow Embassy with that of U.S. personnel of other embassies where many conditions were similar but there was no microwave radiation. Particular attention was paid to measurements on employees at the Moscow Embassy who had been there prior to February 1976, because at that time a special screening had been installed to drop the microwave levels in the embassy to well below 1 μW/cm^2, where not even Soviet work gave any indication of effects. The findings showed that there were indeed higher than normal lymphocyte counts in the employees who were at the Moscow Embassy prior to February 1976. These levels persisted until about August 1977, however, and then dropped off. Employees entering Moscow Embassy service between February 1976 and August 1977 *also* developed on the average higher than normal lymphocyte counts, and these too dropped in August of 1977. The logical conclusion was that these high counts were in no way related to microwave irradiation but resulted from some unknown microbial infection of relatively low level. Such effects are common and not a serious health problem.

In a final attempt to determine whether or not the microwave radiation in the years between 1953 and 1976 had had any measurable effect on the health of the embassy employees a major study was undertaken by the School of Hygiene and Public Health of the Johns Hopkins University under the direction of Dr. Abraham Lilienfeld. The health history of Moscow Embassy employees was compared to the history of U.S. employees of our embassies at other Eastern European locations such as Budapest, Leningrad, Prague, etc. Government medical records, death certificates, medical questionnaire histories, and telephone interviews were used as basic sources. We can look first at the statistics on deaths. Of the 1719 people from the Moscow Embassy for whom adequate information existed 56 had died. Of 2460 employees of the other embassies 138 had died. Both these records were better than normal expectancy, and no significant differences could be seen between them. Similarly an exhaustive study of over 40 different types of illness was also made. The Moscow group had a significantly higher incidence of skin problems (psoriasis, dermatitis, etc.) than the control group, but the incidence was highest in those estimated to have received the least amount of radiation.

The upshot of the whole study including both the employees and their dependents was that there was no evidence of any adverse health effect induced by the microwave irradiation over all these years. If we know the reason for the irradiation it has not been disclosed, so in that sense it is still the Moscow Mystery. There is no longer a mystery about the effect of the irradiation on the employees and their dependents, however. There was no effect. This is helpful in establishing a lower limit on amounts of microwaves that can cause discernible harm, since the maximum levels seem to have been about 18 μW/cm^2, but it does not help us very much concerning the effects of intensities from 50 to 1000 μW/cm^2. The *Saratoga* experiments indicated no effects below 1000 μW/cm^2, and the monkey studies indicated some possible effects between

1000 and 5000 μW/cm² (1 to 5 mW/cm²), but neither study was very conclusive.

There is perhaps one other lesson that can be learned from the Moscow Mystery. It is recognition of the ease with which suspicion can lead to exaggeration and exaggeration can lead to distortion until the truth becomes obscured. This is particularly true when concealment of the true reason for tests and studies is involved. When such distortion occurs it is a long and painstaking process to develop sufficient facts to permit truth and objectivity to emerge again. And for those viewing the process from the outside there is the question of how to recognize it when it does emerge. For this the only guide is the integrity of the scientists making the studies, the committees reviewing them, and the people reporting them. Open, objective studies are the only path to truth.

QUESTION OF STANDARDS. The growing understanding of the harmful effects of low levels of environmental pollution, both chemical and radiative, should properly be reflected in realistic standards and realistic controls to ensure that the standards are met. Seldom have national standards differed as much as those established by the USSR on the one hand and the United States on the other in respect to microwave radiation. While the basis for these differences in standards is gradually becoming clearer, the standards themselves have not yet changed to reflect that clearer understanding. We cannot leave the discussion of the effects of microwaves without further consideration of that issue. During the 1960s and 1970s the permissible standard for continuous occupational exposure for one day in the USSR was 0.01 mW/cm² (10 μW/cm²). In the United States for reasons previously recounted it was 10 mW/cm², a factor of 1000 greater. To talk intelligently about who is right or who is wrong we will have to look at least briefly at how standards are generated in areas such as this and how, once generated, they are policed.

In the United States we have a multitude of mixed and overlapping institutions for the setting of standards. We have the various governmental agencies such as the Environmental Protection Agency (EPA), the National Institute of Occupational Health and Safety (NIOSH), The Bureau of Radiological Health (BRH) (part of the Food and Drug Administration), the Consumer Product Safety Commission (CPSC), and the Occupational Safety and Health Administration (OSHA). We have semiindependent bodies such as the National Council on Radiation Protection (NCRP), a nonprofit organization chartered by Congress in 1964, which "seeks to promulgate information and recommendations based on leading scientific judgement on matters of radiation protection and measurement and to foster cooperation among organizations concerned with these matters." NCRP works through a multitude of committees manned by more than 400 scientists who provide their part-time services on a voluntary basis. We have a variety of industrial standards groups such as the National Electrical Manufacturers Association (NEMA) and the American National Standards Institute (ANSI), which sets nonmandatory but almost universally acknowledged industrial standards in many areas. There is also a host of professional technical societies wielding considerable influence in setting standards and distributing information about them in the technical areas in which they are focused. Examples relevant to our discussion on microwaves are the Radiation Research Society and the Health Physics Society. Finally there are international groups such as the World Health Organization and the International Radiation Protection Association (IRPA), which take on particular importance when wide differences in views exist between different parts of the world as is true in the microwave controversy.

With all these diverse groups having an interest in the establishment and maintenance of standards, it is not surprising that the emergence of a universally accepted set of standards in the United States is a slow and sometimes painful procedure. The balance in the points of view of each group will be different. The Health Physics Society may lean on the side of absolute safety regardless of cost, while the NEMA may have greater concern with the economic aspects of achieving an acceptable safety level. The government agencies, which in the final analysis can set the only truly binding standards, must try to find a middle ground amidst all the pressures. The use of part-time volunteers on important committees of the NCRP and on the advisory committees of the government agencies tends to slow down an already complicated process for reaching a consensus. Nevertheless, this relatively slow and unwieldy process results in ample forums for the presentation of all sides of complex standards disputes. It assures the attention of competent scientists and engineers familiar with all aspects of the issues. The many government hearings set up by the executive branch and by Congress provide opportunities for public-interest groups to establish their viewpoints also. It is these processes that are now under way, trying to assimilate the old and new information on microwave effects from the USSR and its satellites and the new information confirming or refuting it from U.S. laboratories.

Some changes have already resulted from this scrutiny. In 1982 the ANSI issued a modification to their microwave standard and identified it as C95.1–1982. It lowered the maximum permissible irradiation level from 10 mW/cm² to 1 mW/cm² in the most sensitive frequency range, that between 30 and 300 MHz. Between 1 and 100 GHz the level was lowered to 5 mW/cm² and a sliding scale was permitted between 300 MHz and 1 GHz on the assumption that the higher the frequency the less the hazard in this range. The EPA is proposing an even lower limit (100 μW/cm²) in the most sensitive frequency range, but as of this writing no final decision has been made.

The USSR, on the other hand, has proposed very modest *increases* in the permissible levels in their standards. In the most critical range from 30 to 300 MHz this is only an increase from 2 μW/cm² to 3 μW/cm², so there is still a large gulf between the U.S. and the USSR standards and any proposed modifications. It seems probable that continued dialogue will result in gradual further changes towards an accepted standard that will be truly international.

In spite of the differences there is little evidence that the standards applied in the past have permitted appreciable harm to either workers or the public. Most documented injuries seem to have come from exposures to intensities above 10 mW/cm²; thus these are above even the most lenient standards. The current reductions are clearly prudent, however, and if continuing research develops hard data on significant blood chemistry or nervous system responses below these levels, new lower standards will certainly result. The philosophical question is whether we should pay the cost of lower standards now, or maintain the higher permissible levels with the chance that later studies will prove them capable of having caused unpleasant but not life-threatening side effects. When science is practiced as it should be, with free intercommunication of data and ideas, a semblance of truth eventually emerges. It will not be agreed to by all, nor should it be, but it will function effectively for man's benefit.

RADIO WAVES AND HIGH VOLTAGE TRANSMISSION LINES

The final areas to be considered under the effects of nonionizing radiation are long radio waves and high-voltage transmission lines. These can be considered together because the effects of both come from the electric and magnetic fields associated with them rather than from true electromagnetic waves. In the case of very long radio waves, the waves themselves are completely reflected from an object as small (relatively) as a human body, and no effects are possible unless the powers involved are so high as to generate appreciable electric and magnetic fields in the vicinity of the radiating antenna. With extra-high voltage electrical power transmission from 365,000 volts (365 kV) and up, the 50 or 60 cycle waves are contained within the transmission wire, but there are two laws of physics that result in fields surrounding the wires carrying the current. The first law states that wherever there is an electric charge there will be an electric field around it. The term "field" merely means a region in which another electric charge will experience a mechanical force on it when it is in the field. The field strength varies as the magnitude of the initial charge varies. At a given point it will be larger if the charge is larger, smaller if the charge is smaller. The field strength also varies with distance, but in mathematical terms as the square of the distance. The further away one moves from the electric charge, the weaker is the field. At 2 feet from a charge, for example, the field is one-fourth what it is at a foot away, and at 3 feet it is one-ninth as strong. It thus drops off rapidly with distance.

The second law involves the fact that a *moving* charge creates a *magnetic* field about it. This magnetic field will exert a mechanical force on any magnetic material placed within it. More important to our discussion is that if an electrical conductor is moved in a magnetic field, it will generate an electric current in the conductor. This is how an electric generator works, so it is very important to us. In our present discussion it is important because the human body is a reasonably good conductor of electricity; and when a person moves in a high magnetic field, tiny currents can be induced that conceivably could have subtle effects on nerve and brain functions.

Our concern with the effects of very long radio waves and high-voltage lines is thus a concern over possible effects from the electric and magnetic fields associated with them. Since the strength of these fields drops off rapidly with distance we are usually only concerned with effects relatively close to the source. Let's look at each type of field separately and see what kind of interactions can occur, and let's start with the electric field as it surrounds high-voltage transmission lines.

We must first remember that high-voltage transmission can be either direct current (DC) or alternating current (AC). The field around a DC line is constant in sign, positive or negative, and usually constant in amount at any given point. Around an AC line, however, the field is changing in amount and sign, following the 60 cycle per second changes in the alternating current flowing in the transmission line. If an electrical conductor is located in a changing electric field, a current will flow in that conductor. We pointed out earlier that one's body is a conductor, and very small electrical currents can be induced in it by a changing electric or magnetic field. It is thus clear that a person standing close to a high-voltage AC transmission line will have tiny electric currents caused to flow in the various conducting circuits of his or her body. The question is whether these currents are significant enough to cause any physical or psychological harm. We can show that in the case of transmission lines the currents from electric field effects are greater than those from magnetic field effects. Moreover we have pointed out above that AC lines, because of the alternating fields associated with them, have greater effects than DC lines. We can simplify the discussion, therefore, by limiting it to the electric field effects from high-voltage AC lines.

To get the discussion in perspective we first need to compare the magnitude of the fields from transmission lines with the magnitude of other naturally occurring and man-made fields. The strength of an electric field is measured by the voltage difference existing across a given distance, i.e. the voltage gradient. It is usually expressed as volts per meter (V/m) or thousands of volts per meter (kV/m). For very high voltage transmission lines the maximum field on the ground under the lines will depend on the voltage, the height of the lines above the ground, and the spacing between the lines. Typical values for lines from 345 to 765 kV (the highest voltage now in commercial use) will be 5 to 10 kV/m. The normal DC field outdoors in fair weather is 120 to 150 V/m, but in a thunderstorm values as high as 10 kV/m are not unusual. The electric field intensity (AC) one foot under an electric blanket is about 200 V/m, but because of its relatively close coupling with the human body it can induce currents equivalent to those induced in someone standing erect in a field of about 1 kV/m, or one-tenth the field directly under a 765 kV transmission line. Certain electrical appliances such as electric lawnmowers, hedge trimmers, or drills, particularly older models, have fields associated with them up to 33 kV/m, while pacemakers used to regulate heartbeat cause local body currents equivalent to even higher fields. Thus the maximum fields associated with high-voltage transmission lines are not unique in our radiant world, nor are they in ranges where harmful effects have been observed from other sources.

We cannot dismiss high or even moderate levels of electric or magnetic fields as being completely without effect on humans and animals. While documented cases of significant harm are rare, there is a background of measured effects and a few instances of claimed harm giving us pause and requiring further examination. One of the most important cases involves a group of workers in an electrical substation in the USSR where 45 workers were exposed for up to several years to 50 Hz electric fields of up to 14.5 kV/m during much of their working day. There were numerous complaints of headaches, sluggishness, and fatigue, and medical examination showed slightly lowered response time on a number of tests involving nerve and muscle response. Electrocardiogram studies of heart function and studies of blood chemistry also showed numerous abnormalities. The study did not include an adequate control group of nonexposed workers, however, and there did not appear to be evidence of permanent damage. The studies resulted in installation of electrical shielding around much of the equipment to reduce the fields to below 5 kV/m wherever possible. Regulations in the USSR call for limiting the periods of permissible exposure to electric fields between 5 and 25 kV/m and they prohibit exposure above 25 kV/m.

Two Spanish studies have also reported abnormal findings on electrical switchyard workers where high but unspecified electric fields existed. The reported symptoms included headaches, fatigue, and digestive problems, but all the problems appeared to be temporary. In contrast, a 1978 Swedish study with a carefully monitored control group showed no medical or psychological differences between the controls and a group of 53 workers exposed to electric fields in a 400 kV electric substation. Similarly a University of Toronto study of 30 high-voltage repairmen and switchyard workers showed no difference in medical and psychological tests between the exposed group and a matched group of people not exposed to unusual electric fields. Linemen tested had been employed in high-voltage line maintenance at Ontario Hydro for up to 10 years and on a typical working day would have been exposed to fields above 25 kV/m for 15 minutes and to fields between 5 and 25 kV/m for about 6 hours. The tests showed that there were no measurable cumulative effects of exposure to these levels of electric fields over a protracted period of time. The existence of temporary effects while actually in the high fields was not ruled out, since the medical tests were not made in the field environment itself.

Tests on mice and rats have shown that long-time exposures to 50 Hz AC electric fields of 100 kV/m (5 to 10 times greater than fields under a 765 kV transmission line) cause increases in certain types of white blood cells during exposure but that these counts return to normal within 9 days after removal of the field. There is also evidence of slight temporary changes in blood chemistry in humans exposed to electric fields from 1 to 20 kV/m.

There have been many other studies of the effects of electric fields on people, animals, and tissue samples. As is so often the case with measurements made just at the edge of where effects begin to be noted, the data in ranges of interest to us are unclear and often conflicting. The broadest conclusions are that AC electric fields appear to be harmless below about 1 kV/m and are probably incapable of causing permanent harm even for relatively long-time exposures up to 100 kV/m. In the range between 1 kV/m and 100 kV/m it is evident that temporary effects can occur including minor blood chemistry changes and mild psychological effects such as headaches and slight impairment of nerve and muscle function. The USSR regulations limiting the time of exposure at levels of from 5 to 25 kV/m appear to be reasonable precautions, but even exceeding these limits appreciably seems unlikely to cause drastic permanent damage.

These appear to be safe generalizations for 50 to 60 Hz AC fields such as are commonly associated with electric power transmission. There is one slight flag of additional caution that must be raised when fields in the frequency range of 1 to 10 Hz are considered. This is the frequency range of human brain-wave currents, and there is some evidence from tests on monkeys that fields as low as 1 to 100 V/m at these frequencies can cause small changes in the speed and precision with which the brain and nerves respond to external stimuli when the field is on. Studies of tissues from chick and cat brains also indicated that low fields of these frequencies could alter the rate of movement of calcium and GABA, a complex organic chemical molecule involved in transmission of nerve signals. It is conceivable, but far from proven, that there are particular frequencies near those our brain uses to send its signals that have maximum capability for affecting our mental dexterity and function. There is no evidence that these frequencies can cause permanent harm, only that they might influence behavior in the field itself. Even though these frequencies are not in common commercial or industrial use, this is an area where research is needed to develop more precise information on the effects that can occur, the frequencies that are most important, and the threshold intensities where effects first appear.

The effects of *magnetic* fields likely to be encountered in daily living appear to be insignificant to humans. We measure magnetic fields in units of gauss, and the natural magnetic field of the earth is about 0.25 gauss, although this will vary somewhat depending on where you are located on the globe. The added magnetic field directly under a high-voltage transmission line is about the same amount and appears to have no known effect on us. Some household appliances, however, have much larger magnetic fields in regions close to them. Magnetic fields near a color TV receiver may be 1 gauss (1 G) or more and fields near a hair drier or a soldering gun may exceed 10 G. Even an electric shaver can have a magnetic field of 5 to 10 G around it, but much larger fields than this still appear relatively harmless to humans. In a wonderful new medical diagnostic tool known as NMR (nuclear magnetic resonance) advantage is taken of the highly selective magnetic interaction of hydrogen atoms with radio frequency radiation in the megaherz range. This permits computerized scans from which body cross-section images can be formed, often superior to X-ray scans. NMR requires the patient to be in magnetic fields of several thousand gauss. There is no evidence that even these fields cause permanent biological side effects for short exposures, and NMR currently appears safer than X rays. It is a new technique, and its increasing use emphasizes the need for greater knowledge about possible subtle effects of such high magnetic fields.

These subtle effects do indeed exist as numerous studies have already shown. For the most part these effects begin at fields above 200 G, and both beneficial and harmful effects have been observed. On the good side, faster healing of broken bones has been observed in rats exposed to 500 G magnetic fields, and fields of 4000 G have caused reduction in the size of cancerous tumors in mice. On the negative side, fields from 100 to 4000 G have resulted in interference with white blood cell formation in mice, subtle changes in the biochemistry of frog and mouse tissues, temporary effects on guinea pig lungs (200 G), behavioral changes in rats and monkeys, etc. All these experiments, however, leave one with the impression that compared to other environmental hazards magnetic fields are relatively benign. The advent of NMR diagnostics, requiring short-time exposure of humans to very high magnetic fields, will certainly bring about greatly increased research on this topic, but present indications are that the risks are small.

Organisms other than man sometimes show extraordinary sensitivity to the presence of even very small magnetic fields. It is now believed, for example, that birds and fish rely at least in part on their sensitivity to magnetic and electric fields that provide navigational guidance for migration over thousands of miles. Certain bacteria contain tiny particles of a magnetic iron oxide (Fe_2O_3) and use the magnetic interaction between these particles and the earth's magnetic field to guide them towards the bottom of ponds or oceans in which they exist. Their direction of motion can be changed by applying weak directional magnetic fields to their environment.

There are many other examples in nature of beneficial utilization of the natural magnetic and electric fields pervading our environment. Man's generation of artificial fields is almost always over limited areas and seldom intrudes on these sensitive adaptations. Nevertheless consideration of the environmental impact of new ventures changing the nature of our radiant world should always include the possible disruption of organisms whose behavior may be keyed to the levels of electric and magnetic fields that have existed naturally in our world for ages.

PART 4

RADIATION
AND YOU

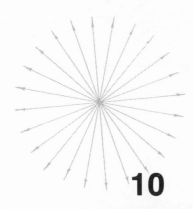

10

RADIATION IN OUR DAILY LIFE

Now that we have learned what radiation is, where it comes from, and what it does to us, we can consider sensibly the many issues we face in our society in which radiation and its effects are a key factor. To what extent should radiation be used in medicine? Is there really a dangerous "electronic smog" from microwaves? Should the development of ultra-high voltage electric power transmission be continued? Are there radiation hazards in our home or workplace? What should be done about nuclear weapons? What should be done about nuclear power? In almost all areas we will have the difficult problem of balancing a risk against a reward. We do not all see the balance alike and hence we may come to different conclusions. Perhaps we can at least see more clearly what the balance really is and arrive at our conclusions from the basis of reason rather than untutored emotion.

IN MEDICINE

We will start with an area where the final conclusion is almost universally accepted. The use of radiation in medicine has brought enormous benefits to mankind. It goes back to the early history of X rays when the ability to see breaks or deformities in bones opened a new era in the treatment of fractures. The diagnostic power of X rays has evolved steadily since their discovery in 1895, until today we have the complex computerized tomographic devices creating pictorial cross sections of the entire body including the brain, giving doctors tremendous insight into the effects heredity or disease have introduced. The use of radioisotopes as a diagnostic tool is more recent and is truly a by-product of nuclear power that created the many radioisotopes in sufficient quantities to permit their universal use at relatively low cost. We have mentioned the use of iodine isotopes to determine thyroid function, the use of technetium-99m for liver and brain scans, and ^{51}Cr for blood chemistry studies. Many other isotopes are used also in increasingly sophisticated and specialized diagnostic tests.

Most of the isotopes used come from nuclear reactors designed for both research and isotope production. Complex chemistry is required for the separation of these isotopes in pure form, and the separation leaves behind a radioactive waste of isotopes whose characteristics are not suitable for medical or industrial use. The benefit of the diagnostic isotopes is thus tempered by the need for disposal of a hazardous waste. Even the diagnostic isotopes themselves, although usually of short half-life, require careful handling both of the material before use and of the excretions from the patients on whom the isotopes are used. The amounts of waste generated by or associated with medical use are far smaller than those produced by nuclear power, but they are a part of the problem of disposal of nuclear wastes. They are almost all classified as low-level radioactive waste, but there are only three approved low-level waste dumps in the United States, one in the state of Washington, one in South Carolina, and one in Nevada. All hospital radioactive wastes must be shipped, at considerable expense (about $250 per barrel) to one of these sites. Yet there have been several occasions, particularly in 1980 to 1981, when these sites were closed to all but local wastes, posing a considerable nuisance for hospitals across the country. The Low-Level Radioactive Waste Policy Act of 1980 established the policy that each state should be responsible for disposal of its own low-level radioactive waste either through compacts with other states for a common facility or through their own facility. The states were given until 1986 to establish such an approved site, and at that time the three operating sites mentioned above could opt to refuse waste from other states not having a definite agreement with them. Many states have not moved rapidly enough to arrange for individual or common use sites within this deadline, however. A pending crisis in low-level radioactive waste disposal was averted by passage in late December 1985 of the Low-Level Radioactive Waste Amendments Act of 1985. Under this act every state must have ratified legislation to set up a waste disposal compact with adjoining states or must agree to build its own repository. States failing to comply with a complex set of deadlines must pay continually escalating charges for use of the existing operating sites. Any state failing by 1

January 1996 to be a member of a compact with an operating site or to have a site of its own will be barred from use of the other national sites but will be required to take title to all low-level radioactive waste produced in the state. The onus of disposal will therefore be transferred from the generator to the state, and the state will have no disposal location available—an unenviable position. The federal government is thus forcing the states to act to provide adequate disposal facilities at the earliest possible time. The technical problem of disposal of low-level wastes is not a difficult one, but each citizen should make sure that his or her state promptly establishes the necessary regional agreements or has plans to provide its own facilities. Thirty-seven states have already done so. The benefits through medicine alone are enormous and the risks trivial if sound engineering is employed.

Both γ rays, usually from ^{60}Co, and X rays are used for treatment of cancer. The radiation must be carefully focused so as to give maximum dose to the cancer area and as little as possible to the healthy tissue surrounding it. It is ironic that the very tool that can destroy cancer is the same one known to be one of its causes. The possibility of side effects later on thus always exists with radiation treatments, but when cancer is the present enemy the doctors can gain the reward of life now for the price of only possible complications long in the future.

There is no question but that the use of radiation for diagnosis and treatment of disease has been of enormous benefit to man and will continue to be as the doctors and radiologists learn more about how to get better diagnostic pictures with lower levels of dose, how to combine radiation and chemotherapy in cancer treatment, and how to focus and control more precisely the powerful X-ray beams. Particularly in the diagnostic area there is real and proper concern about how much is enough. Is a dental X ray every 6 months really a good idea or is it more reasonable to defer to once a year or even less often? There are conflicting trends in this issue. On the one hand with better films and more precisely focused X-ray beams, the exposure per film is being steadily reduced. On the other hand there is the uncertainty that still exists about the statistics of cancer causation by low levels of radiation. In this case the risk is very, very small, but the gain in information by increased frequency of X-raying is also small. It is wise to discuss this issue with your dentist to make sure that both he and you understand the characteristics of his machine and its use and the extent of benefit from the schedule of X rays he plans.

For many years prior to 1970 it was common to use X rays to determine the position of a foetus within a pregnant woman so as to be better prepared for delivery. Usually this would be late in the pregnancy, but occasionally it would be relatively early. As greater understanding was developed concerning the effects of low levels of radiation, studies were undertaken to see whether the practice was really safe. Two studies in particular, one by Drs. Stewart and Kneale in England and the other by Dr. MacMahon in the United States, showed that diagnostic radiation of pregnant mothers led to an increase in early deaths from cancer (before 10 years old) for their children. Such early cancer deaths are not common, however, and the increase was not very large. But estimates seem to show that if a million pregnant women received 1 rem of diagnostic X-radiation (about 4 to 5 pictures each), over a period of 10 years there would be about 500 extra cancer deaths in their children. There is other evidence that indicates this estimate may be high, but it seems safer for the moment to assume it is real. It thus says that there is a 1 in 2000 chance that multiple X rays during a pregnancy will cause the death of the child by cancer some time in his or her first 10 years. While this is not a high probability, it is not negligible either. It says that such X rays should be avoided unless there is a clear overriding medical need that would warrant this risk. As a result there has been a substantial drop in the use of X rays during pregnancy, particularly in the first 3 months of pregnancy when the risk is apparently highest.

In general it is a wise precaution to minimize total X-ray exposure during one's lifetime. This is particularly true in the early years of life because of the apparently long incubation time over which cancer effects may develop. Every person should keep a record of all diagnostic and other X-ray treatments received so that this record can be available to doctors and radiologists who may have to decide on the desirability of a specific set of X-ray tests. You should have no fear or concern over X-ray or radioisotope tests or treatment prescribed by a competent doctor who knows very well the risk/reward ratios involved. It makes no sense, however, to needlessly duplicate X-ray tests when a recently taken film may serve as well. Radiation in medicine will almost certainly help every one of us sometime in our lives. Remember its small risk and treat it wisely.

IN THE HOME

When we think about controlling the environment in our home we usually think about temperature, humidity, and sometimes the chemical purity of the air. We have furnaces, stoves, or heaters to keep our houses warm in the winter and air conditioners to keep them cool in summer. Some of us use humidifiers or dehumidifiers to make the atmosphere inside moist or dry as we prefer or our climate makes necessary. We have exhaust fans to eliminate cooking smells from our kitchens, chimneys to take out furnace fumes or fireplace smoke, and we often open doors or windows to air out the house from bug sprays or cleaning solutions.

These are conscious attempts to control the home environment. We turn on a light, however, without thinking that by doing so we are changing our radiation environment. Light and darkness are so much a part of our everyday life that our minds treat them as separate and familiar phenomena. We don't think of them in terms of presence or absence of electromagnetic radiation of a special frequency range to which our eyes are attuned. That is what they are, just as the sensation of heat or temperature is merely the body's reaction to the balance of electromagnetic radiation in the thermal range that it is emitting to other objects such as cold walls or receiving from them as in the case of radiators and fireplaces. You have probably seen ads for quartz-tube–radiant-heating units touted for their high efficiency because they "heat the people in the room, not the room itself." The trick is to have a high temperature, electrically heated core rod with a reflector behind it focusing the infrared radiation towards the individual seeking comfort. The efficiency of conversion of electricity to heat is no greater than in a lower temperature unit circulating the heat into the room by blowing air past the heating unit and into the room. The difference is that in the quartz heater a greater percentage of the heat generated is utilized directly in giving a sense of warmth to the person on whom the unit is directed. When the unit is turned off, the room will immediately feel cold again because less of the energy has been used to warm the air, the walls, and other objects in the room than in the case of the circulating fan heater.

This leads us gradually toward other less familiar and sometimes potentially hazardous aspects of radiation in the home. Extending from the visible light spectrum at the high-frequency end is the region of ultraviolet or UV, which we cannot see but includes the frequencies most effective in causing the skin pigment changes known as tanning. Many people use sunlamps to maintain an attractive tan or because they believe there are other health benefits from regular exposure to the UV component of sunlight normally missing from incandescent light bulbs. Properly used these lamps may have significant benefits, but there are risk/reward considerations in their use. The painful reddening of the skin, which we know as sunburn, is largely caused by the very narrow range of wavelengths from 290 to 320 nanometers within the ultraviolet range. A broader range of frequencies is involved in the tanning process resulting from the light-induced formation of a substance in the skin called melanin. The melanin darkens the skin and protects against the sunburn response. Cautious and gradual use of sunlamps can promote tanning and avoid sunburn. Continued use, however, over long periods may increase the probability of skin cancer just as does prolonged exposure to sunlight.

There are other subtle effects of light that apply indoors as well as outdoors. It has been found that the rate of formation of the important vitamin D_3 is dependent on the absorption of ultraviolet radiation by the skin. This vitamin influences the rate of absorption of the element calcium, an essential in bone formation. While most of us get enough natural sunlight for this function, the use of sunlamps can be helpful, particularly when a person is housebound for any reason for long periods. Incandescent and normal fluorescent lamps do not include enough of the ultraviolet region in their spectrum to provide assistance in vitamin D_3 generation. Special fluorescent lamps can be produced that will extend their spectrum into the ultraviolet region. This suggests that in lighting design we should consider more seriously the effects beyond those that are solely visual, i.e., sufficient intensity for adequate vision and broad enough spectrum of wavelengths to provide adequate color rendition.

In such considerations we have to weigh the good and the bad. There are other tests showing that genetic damage can be produced in certain cells by irradiation in the UV range. Are these data relevant to our everyday use of light? Most probably they are not, but they indicate the complexity of the irradiation effects that can occur and the fascinating balance of benefit and harm that must be considered with all types of radiation. Lighting in the home is man's servant and benefactor. Even it requires sensible use, and even it can be made more useful by continuing to increase our understanding of its subtle effects.

Another casually accepted feature of the American home is the television set. Its picture is formed by an electron beam darting across the light-emitting phosphors on the inner surface of the tube. The electrons themselves cannot penetrate the glass front of the tube, but at high enough voltages when striking the right materials they can produce X rays that have high-penetrating power and can emerge from the set into the room. The old black and white TVs operated at relatively low voltages, and X-ray emissions from them were trivial. With the advent of color television sets in the late 1950s, however, the problem became more severe. These sets used higher voltages to produce the color picture, and for the first time X-ray emission became a problem. In 1955 the International Commission on Radiological Protection (ICRP) recommended that emissions of X rays from TV sets should not exceed 2.1 milliroentgens per hour (mR/h) "at any accessible surface." A field survey in Florida in 1968 showed that of 149 sets checked at random, 23 exceeded the recommended levels and 2 were as high as 100 mR/h. Findings such as these prompted further investigation of the problem that disclosed repair servicemen, to improve a poor picture in an old set, occasionally boosted the tube voltage, resulting in substantial additional X-ray emission.

In 1968 Congress passed a law, the Radiation Control for Health and Safety Act, under which standards could be enforced by the Food and Drug Administration (FDA). The set of standards established in 1970 provided for a maximum exposure rate of 0.5 mR/h at the surface of the set, but with measurement method requirements that actually ensured normal exposure rates would be much less than this. Since then, X-ray emission levels from TV sets have been generally under good control. Improved solid state circuitry in modern sets has helped greatly, and for the most part the only concern today is with older sets, particularly where improper service repair methods may have been used. A few simple precautions will ensure safe viewing:

Buy sets made by reputable manufacturers and approved by the appropriate government agencies.

Use only fully qualified and responsible repairmen and make sure that what they do cannot affect the X-ray emission levels of the set.

Make sure in buying a second-hand set that it has been properly cared for.

Don't stay closer to your TV set than is necessary for normal comfortable viewing.

With these precautions you can be sure the average significant X-ray dose from your TV set won't exceed a few millirem per year, perhaps 1% of that from natural radiation.

A more recent source of radiation in the home is the microwave oven. Its function requires it to operate in a microwave frequency range that easily penetrates and is absorbed in human tissue just as it is in the food to be cooked. It is important, therefore, that the radiation be contained within the oven itself and not permitted to leak out into the kitchen. This is accomplished by carefully designed seals to ensure that the ovens as manufactured meet the BRH standard of emissions no greater than 1 milliwatt per square centimeter (mW/cm^2) measured at any point 5 centimeters (about 2 inches) from the oven. After installation and use the emissions may not exceed 5 mW/cm^2 at the same point. This allows for expected normal wear on the seals and is the maximum amount expected in normal home use.

You will recall, however, our discussion of the difference in acceptable microwave standards in the USSR compared to the United States, and you will probaby note that the *emission* number mentioned above is not far below the recommended *U.S. exposure* limit of 5 to 10 mW/cm² for continuous exposure. Here we have to emphasize again the difference between the terms *emission* and *exposure*. If you stayed just 2 inches away from your oven all the time it was on you would receive a maximum exposure level of 5 mW/cm², which is the maximum permitted emission level at that point. In practice you are seldom if ever just 2 inches from the oven, and since the power density decreases rapidly with distance your actual exposure during normal cooking will be very much less than the permitted maximum emission level. Theoretically if an oven is leaking microwave radiation at the maximum allowed rate, the exposure level at 3 feet from the oven will be 0.02 mW/cm² (20 μW/cm²) and at 6 feet it will be 0.005 mW/cm² (5 μW/cm²). These are safe levels even under the highly conservative USSR standards.

When a housewife is putting something into a microwave oven to heat for only a minute or so she may remain relatively close to the oven waiting for the bun to heat, for example. If a roast is in the oven for a half hour or so she will move about the kitchen doing other things and will on the average be farther from the oven. The typical pattern will therefore be somewhat higher exposures for short periods, considerably lower levels for longer periods. Since any subtle effects of microwave radiation are a function of both exposure level and exposure time, the habit patterns of the housewife tend to increase her safety by decreasing the average level of exposure when longer times are involved.

When all these factors are added up it can be concluded that a well-engineered and well-built microwave oven is a safe and sensible addition to the kitchen even when our imperfect knowledge of the subtle effects of low levels of microwaves is factored into the risk/reward consideration. It should not be taken wholly for granted, however, and a few sensible precautions are worth observing. They include:

Buy reputable microwave ovens made in accordance with Bureau of Radiological Health standards.

Take good care of the oven door seals and make sure that any needed repairs are made by authorized and competent service people.

Don't buy used old microwave ovens (particularly ones made before 1971) unless they have been carefully overhauled and tested by a reputable repair service.

Don't operate an oven if the door doesn't close properly or is bent or warped.

Don't let children sit or stand directly against the oven for any long period while it is operating, and by the same token use comfortable care yourself not to stay too close too long.

Over the past 10 years there have been many improvements in leakage control by the manufacturers of microwave ovens. These plus sensible use as described above can make a microwave oven a safe, energy-efficient convenience in any modern kitchen. So good eating and good health!

Another source of radiation in the home is the smoke detector. These highly effective safety devices operate through the monitoring of a very tiny current generated by a battery and flowing through a small column of air ionized by radiation from either radium-226 or americium-241. When smoke enters the device the electrical conductivity of the ionized air column is changed, and it is this change from the normal level that activates the alarm. ^{226}Ra has a 1600 year half-life and gives off 4 MeV α particles. ^{241}Am has a half-life of 458 years and emits 5 MeV α. The americium source is now more commonly used than radium, and a typical unit will contain 50 μCi or less of ^{241}Am. This small amount plus its being an α emitter in a sealed source means that the radiation exposure in the home is insignificant even when cleaning or testing is involved. It is not safe, however, to try to take the units apart in such a way as to damage the sealed radioactive source. Care should be taken also in disposal of old units, and some manufacturers recommend return to them if disposal is necessary.

We come now to one of the most complex and least understood parts of radiation exposure in the home. It is the topic of radon which, unknown to most of us, is probably our major source of home exposure to radiation. Radon is a radioactive gas in the long decay chain that begins with uranium. Because uranium is a common mineral, radon is also common, and being a gas it can spread easily over large areas. Its presence can only be detected by sensitive instruments and it can be, and probably is, in very small quantities in the air you are breathing now. Our true concern is less with the gas radon itself than with the solid daughter products, particularly the α particle emitters polonium-218 and polonium-214 resulting from the radioactive decay of radon (^{222}Rn). The half-life of radon is only 3.8 days, so it is decaying relatively rapidly as it enters and leaves our lungs. It decays by emission of an α particle and is converted to polonium-218 (^{218}Po). The ^{218}Po has a half-life of only 3 minutes and decays by emission of another α particle to lead-214 (^{214}Pb) which has a half-life of 27 minutes and converts by way of bismuth-214

(^{214}Bi) emitting β and γ with a 20-minute half-life to become 214 Po. This polonium isotope exists for only a fraction of a second and converts to ^{210}Pb by emission of another α particle. The rates of decay slow down at this point since ^{210}Pb has a half-life of 19.4 years. It decays by β emission to 5-day bismuth-210 (^{210}Bi) that yields ^{210}Po, another α emitter with a 138 day half-life. The product of this α emission decay is lead-206, which is a stable isotope and the end of the uranium decay chain.

This is a very complicated decay chain, but there are really only two qualitative features you need to understand in relation to the hazards of radon. First, it is the α particle emissions far more than the β particle emissions that are important, and then only when the emitting isotopes are actually present in the lung. Thus it is the three α-emitting polonium isotopes ^{218}Po, ^{214}Po, and ^{210}Po that are the most serious hazards. Second, it is not just the decay from radon within the lung that is important, but the room environment itself as well. This is because in a dusty, stagnant air environment the polonium daughter products and their lead and bismuth precursors can, as they form, attach to floating dust particles or settle on walls and floors where they can be stirred up and breathed into the lungs separately from the radon gas itself. Thus the level of these dangerous daughter products in a poorly ventilated space may be either higher or lower than the amount normally in equilibrium with the ever-present radon gas.

The equilibrium amount of any radioactive isotope in a decay chain is determined by its half-life and the half-lives of the isotopes preceding it. In reaching equilibrium the amount of any given isotope will increase until its rate of decay just equals its rate of formation. This means that the isotopes of long half-life will be present in larger quantities than the ones with short half-lives. You can perhaps understand this better if you think of a brook flowing down a hillside. If a pool has a very small opening at its lower end, the pool will grow in size until the added pressure and the added areas for outflow permit the outflow to equal the inflow. The concept is the same for radioactive isotopes. Thus in the decay chain from ^{222}Rn the amounts of 19.4-year ^{210}Pb and 138-day ^{210}Po present will tend to be larger than the amounts of the other isotopes. As in the brook where the amount of water flowing out was the same regardless of the size of the pool, so at equilibrium each α-emitting isotope will be contributing equal amounts of α particles—the smaller amounts of short half-life isotopes decaying more rapidly, the larger amounts of long half-life isotopes decaying more slowly.

Our objective in the home should be to minimize gaseous radon and its daughter products as much as possible. To do this we must first know what the sources of radon are. Since it originates from the decay of uranium, we can start by seeing where uranium itself is located. Uranium is commoner than silver or gold or mercury but less common than lead or copper, for example. It is spread very widely in dilute amounts, however, and averages 4.7 parts per million (ppm) in granite, 3.7 in shale, 2.2 in limestone, and 0.5 in sandstone. Wherever uranium is present, radon is present also; it is constantly coming out of the ground at an average rate of about 5 atoms per second for each square inch of ground. The rate will be higher where uranium is prevalent and lower where it is scarce. Typical regions of relatively high radon levels are certain parts of Florida where uranium exists in phosphate sands; portions of the southeastern United States where uranium-bearing shales are common; northeastern United States where granite is common; and regions in Colorado, Montana, Utah, and New Mexico where uranium exists in minable quantities. Total worldwide emission of radon from the ground amounts to about 2.4 billion curies per year. Oceans emit radon too, but at rates 100 times lower than the average over land, so they contribute very little to our total exposure.

There are four basic sources of radon entry into our homes. First is permeation of gas from the ground into basements and then throughout the house. This is probably the largest source and will be greatest in areas where uranium is prevalent in the soil. The amount can be reduced by use of sealants on basement floors and walls, and this is a sensible precaution in regions of high radon concentration. The second source is building materials themselves, particularly where stone or products made from stone are used. In Sweden, for example, an alum shale was often used in formulating concrete for home construction. This shale was relatively high in uranium and contributed significantly to radon levels in thousands of homes. Use of this material has now been banned. A third pathway for radon in the home is through the water supply. If reservoirs or other fresh water sources exist in areas where uranium is present in rock or soil, radon will be present in solution in the water. When this water is drunk it releases radon and its daughter products in our stomachs, and when the water is used for showers or tub baths a portion of its radon is released into the home air. One study has shown that when water is run into a bathtub, 17% of the contained radon escapes into the air within 12 minutes. Measurements in certain areas of Maine where high levels of radon were suspected showed radon in amounts greater than 10,000 picocuries per liter (pCi/l) in 24% of the samples tested and greater than 100,000 pCi/l in 2% of the samples. A fourth source of radon in the home may be a gas stove or space heater. Natural gas emerging from the ground may contain anywhere from 1 to 1500 picocuries of radon per liter of gas. Since the radon in the natural gas is now separated from its solid parent source it will decay during transfer and storage, so the amount present in the gas reaching your home may be much less than this. A series of actual measurements in distribution lines in various cities showed an average of 23 pCi/l. Even this will add a measurable amount of radon to the air in a home where all the cooking is done with gas, however. Gas furnaces usually vent by chimneys to the outside air and thus do not contribute to the radon levels in the home.

This leads us to a consideration of the seriousness of the amounts of radon we have been discussing. To get it into perspective we again have to consider a new pair of measurement units that you will need to get used to. These are the "working level" (WL) and the "working level month" (WLM). Why the need for a new unit? It is because the measurement of quantities of radon alone isn't enough to determine the health hazards. We pointed out earlier that it was the daughter products and particularly the α particle emissions from them that had the greatest potential for biological harm. The WL is defined as "any combination of radon daughters in one liter of air that will result in the ultimate emission of 1.3×10^5 MeV α particle energy." This amount was chosen because it is the α particle energy released by the daughters in equilibrium with 100 pCi of radon gas. The WL unit was developed for measurements in uranium mines where it was first learned how hazardous radon and its daughter products could be. Within these mines, as is sometimes true elsewhere, the amount of daughter products in the air was usually not in balance with the amount of radon. Some of them settled out with dust particles or were otherwise removed, and the total amount present was less than would have been predicted from knowledge of decay rates alone. Where 100 pCi radon was present, therefore, the actual amount of daughter products might be only 0.5 WL rather than the 1.0 WL that should theoretically exist. The important measurement in the mine was the actual measured WL, and in estimating the total risk to a miner, records were kept in WLM where 1 WLM meant exposure to a level of 1 WL of daughter products for a total of 170 hours (the normal working month in a mine).

We continue to use these WL units largely because it was through the high levels of exposure in uranium mines that the lung cancer hazard of radon and its daughter products came to be recognized. Even as early as 1924 an association had been made between radon exposure and lung cancer, but it was not until the 1960s that truly definitive studies showed clearly its extent and the relationship with dose. By then, however, the histories of uranium miners in the United States, Czechoslovakia, and Canada showed clearly that exposure to radon and its daughters was directly linked to later development of lung cancer and that at least for moderate exposures the higher the total WLMs of exposure, the higher was the probability of dying from lung cancer.

To get a feel for the numbers we can consider the study of the cause of death of Canadian miners exposed to varying levels of radon daughters. The study covered miners born in 1933 or before and followed deaths up to 1974. Of those receiving 180 WLM, 4% died of lung cancer. This same disease caused the death of only about 2% receiving 60 to 100 WLM, and less than 1% of those receiving only about 40 WLM. Similarly a U.S. study showed 11 deaths from lung cancer in a group of miners receiving 120 WLM exposure, where only 2 such deaths would have been expected in the absence of radiation. These numbers are not inconsistent with other data we have examined, since 1 WLM is about equal to 5- to 10-rem exposure to the lungs, and miners exposed to 100 WLM would have received 500 to 1000 rem over the period of their exposure.

Average combined indoor and outdoor exposure to radon for a typical individual in the United States has been estimated to be 0.054 WLM/yr or the equivalent of 0.27 to 0.54 rem; so that even over a 70-year life span it is a factor of about 20 lower than the amounts clearly demonstrated to be harmful. We again face the question of whether the effects at low levels received over long times are linearly related to the effects of more concentrated doses. If they are, then some cases of lung cancer in the population are truly ascribable to this radon exposure. The radon level in a typical well-ventilated home in the United States may be from 0.001 to 0.004 WL, which would give an exposure to the occupants of 0.041 to 0.164 WLM/yr, assuming an individual spends 80% of his or her time indoors. Some homes run considerably higher than this, however, and a survey made in central Florida, for example, showed that 15% of the homes measured had levels above 0.03 WL. The extra monitoring for nuclear radiation in eastern Pennsylvania after the accident at Three Mile Island showed that essentially none existed because of the accident. But it was found that a large area of the state, now known as the Reading Prong, had very high levels of naturally occurring radon, some houses measuring as high as 20 to 100 pCi/l, which would mean 0.2 to 1.0 WL if equilibrium existed.

We know that in the Czechoslovakian uranium mines where the average level of radon daughter products was 1 WL, if a miner worked 20 years or more he accumulated an average exposure of 300 WLM. Review of the statistics on deaths in this group of miners showed that the exposure required to double the normal rate of death from lung cancer was 56 WLM. A similar study of Canadian uranium miners showed the exposure to cause a doubling of the rate of deaths from lung cancer to be 17 WLM. Exposure in a home with a daughter product level of 0.03 WL, assuming 80% of an individual's time spent in the home, would accumulate a total of 62 WLM over 50 years of adult life. This number is comparable to the amounts shown to have measurable effects on the uranium miners, although their exposures were at higher levels for shorter times. It is thus possible, but not certain, that continued exposure to radon daughter products in a home at a level of 0.03 WL would at least double the probability of getting lung cancer if one did not smoke. It has been shown that the risk of lung cancer in smokers is *15* times that in nonsmokers, so the decision to smoke or not to smoke is very much more important than the decision whether or not to continue living in a home with a 0.03 WL environment of radon daughter products. There is good evidence from the study of uranium miners that the presence of radon will increase the risk of lung cancer in smokers as well as nonsmokers, so control of radon is desirable whether or not one chooses to smoke.

One of the results of our new found concern about energy has been a tendency to seal up our houses to reduce heat loss. The lower ventilation turnover associated with this can appreciably increase the levels of radon daughter products in the home and thus increase the risk of lung cancer.It is wise to know the levels of radon in your region and consider the type of construction in your home before using drastic measures to seal it up to reduce heat loss. You may be saving money and oil, but the price may include an increase in the risk of lung cancer. This is not a trivial risk either, since one estimate concludes that reduction of the average ventilation level in U.S. homes by a factor of two could lead eventually to 10,000 to 20,000 additional cases of lung cancer per year in the United States. In this country today close to 3% of all deaths are caused by lung cancer. Radon may be responsible for one-tenth of these or 0.3% of all deaths in the United States.

Our home is the one place where we can have significant control over radiation levels. We can make sure that our TV and our microwave oven are of good quality and well maintained. We can use care in the operation of ultraviolet and infrared lamps. We can use proper precautions in the disposal of smoke alarms. We can use adequate ventilation in our homes; and in regions where it is necessary we can take steps to minimize the entry of radon through cellars or out of the walls, gas stoves, or water supplies. We can use radiation wisely for its many benefits, but we should be conscious and knowledgeable about its presence and respectful of it in all its forms.

IN THE EXTERNAL ENVIRONMENT

In the external environment things are different. Even collectively we cannot change the amount of cosmic radiation bombarding the earth, the total amount of radon in the air, or the distribution of naturally occurring radioisotopes such as ^{14}C, ^{40}K, or ^{87}Rb. Individually we cannot control the amount of microwave radiation reaching us from radar installations or from TV transmitting towers. We cannot by ourselves determine whether a 765 kV transmission line should be installed or where it should pass. We can't control alone whether a solar collector in outer space should be installed and radiate its power to earth by microwaves, and we cannot individually decide whether nuclear power should decline or grow. These are societal questions where the risk/reward decisions are made by governmental bodies through the setting of standards and the granting of licenses. The basis for the standards and the authorizations in an ideal world should be full understanding of all of the risks and all of the benefits plus a clear way of balancing these. We do not live in an ideal world, however, and our governments must make these decisions based on the best knowledge available to them and their understanding of how we, their constituents, weigh the balances.

As individuals we cannot advise or elect constructively unless we also understand the risk/reward issues involved. It should be instructive to examine these aspects of technologies generating radiation in our environment, particularly ones that have been areas of public controversy in the past and may continue to be in the future.

WORKPLACE. For many people the major source of radiation is at work. This includes medical radiation technicians, military personnel using radar equipment, nuclear power workers, operators of microwave heat-sealing devices, industrial radiographers, and many others. While workers can minimize exposure by their own care and vigilance, they must also depend on the safety practices of their employers and on the stringency of establishment and enforcement of adequate standards.

We have referred to the complex problem of establishing safe and certain standards, given the type and number of agencies involved. In 1978, for example, a congressional study pointed out that 17 federal agencies alone had at least some radiation health and safety authority. The employer is beset by multiple and sometimes conflicting regulations controlling conditions in his plant.

If the standards are set too rigidly, they may prevent a technology from becoming economically useful. If they are set too loosely, they may result in unnecessary illness or death. Ignorance of the effects of radon in uranium mines permitted lax standards in the early history of uranium mining. The result was an excessive death rate from lung cancer in these workers. Standards must reflect our present knowledge of health effects and be flexible to change either up or down as our knowledge of effects becomes more precise. This accommodation to change in knowledge is slower than desirable in our present system, but it does occur. The debate over proper microwave standards is a case in point.

One of the ethical problems posed by the presence of radiation in the workplace is whether it is reasonable to permit exposure to workers at levels higher than permitted for the general public. There have always been risky occupations—the military, coal mining, high-rise construction, test piloting, stunting, working on offshore oil well rigs, etc. Workers in these fields recognize and accept the associated risks, sometimes for money and sometimes for the challenge. Should working with radiation be considered in this way? In the sense that there is a risk of accidental overexposure when something goes wrong, there is a valid similarity. The situation for radiation exposure is really different, however, because a risk is present even when everything is operating properly and within limits. It is the same risk we all assume just by being alive in the normal radiation environment around us, but very slightly increased by the higher than normal radiation levels of a particular workplace, such as a medical radiation treatment facility. We could not operate the facility if we were required to have absolutely no increased exposure to the operating technicians. But what is an acceptable level of increased risk to the operators in order to obtain the enormous benefits to others that their treatment facility can provide?

There is no absolute answer to this question, and we have to rely on judgement and common sense just as we do in all other types of employment. We set standards for maximum exposure to the public on this basis, and we must use the same judgemental process in the workplace. It is reasonable to insist on larger margins of safety for the general public than for workers in hazardous jobs when these hazards are essentially unavoidable. *Sensible* margins of safety must exist in all industrial standards, and radiation is no exception.

We must now grapple with what we mean by "sensible margins of safety," and this is the hardest problem of all. A fireman entering a burning building to try to save people trapped within may accept very high and obvious risks. This is a single crisis event, however, and the hazard exists for a very short time. Once the job is done the risk is gone. With radiation the risk period is actually decades after the exposure, so the risk is continuing, but it blends into all the other risks we encounter from disease, other environmental exposure, or accident, and there is a very high probability that nothing at all will result.

Studies of Hiroshima and Nagasaki data and many other studies have shown that significant cancer induction effects are only clearly evident after single doses of radiation above 50 rem. Below that level, effects probably exist but cannot be statistically proven because they are too small. We have a natural background radiation exposure of about 100 mrem (0.1 rem)/yr on a continuous basis. National and international bodies such as the National Council on Radiation Protection and Measurement (NCRP) and the International Commission on Radiation Protection (ICRP) have sought to establish limits for man-made radiation exposure above the natural base yet reasonably below the regions of known significant hazard. They have recommended that the dose to any critical organ (such as the liver, kidneys, reproductive organs, etc.) should not exceed 50 rem in any one year for workers and 5 rem/yr for members of the public. Whole body radiation should be limited to 5 rem/yr for workers and 1 rem/yr (10 times background) for members of the public. These dose levels are seen as "the lower boundaries of totally unacceptable levels," and the main recommendation is to keep all exposure levels *as low as reasonably achievable* below these unacceptable levels.

This answers our original ethical question by saying we must keep the exposure of radiation workers less than a factor of 5 times the exposure we permit the general public. Why 5 times? Why not the same, or twice, or 10 times? It is a matter of judgement and choice. Even 5 rem/yr will change the cancer statistics in such a small way that the change will be impossible to determine exactly. It is admittedly a lowering of the safety margin, but as long as we truly strive for levels as low as reasonably achievable it is seen as a valid small risk for very large rewards.

The desirability of attaining levels as low as reasonably achievable (sometimes referred to with the acronym ALARA) raises a number of problems itself. Perhaps foremost is the question of how much an employer is willing to spend to reduce radiation levels already under, but perhaps just under, the recommended limits. It must in the last analysis be a matter of negotiation between regulating bodies, employers, and workers.

There are four primary tools available to an employer for controlling radiation levels where radioactive materials or radiation devices are involved. These are shielding, atmosphere control, procedural control, and monitoring. You have probably noticed that when you have a chest X ray the operator goes into a little cubicle or another room in order to operate the controls. The walls separating the operator from the machine are usually specially designed in thickness or composition (added lead, for example) to ensure minimum passage of radiation through them. The operator, who has to be present all the working day, is thus protected from continued exposure. Similarly the walls surrounding the core of a nuclear reactor are made of concrete many feet thick to cut down radiation leakage into surrounding work areas. How thick and how costly may be questions for negotiation when attempting to achieve levels as low as reasonably possible.

Atmosphere control may also be required, particularly where radioactive materials are involved that can be airborne as dust or gases. The specific requirements will depend on the materials handled, their amounts, and their types of radioactivity. Strict procedural controls are also essential, for example, to avoid spills of radioactive materials or leaks or inadvertent unshielded use of radioactive sources. There must be strict accountability for all radioactive materials to prevent loss or misuse. Finally there must be a high-quality monitoring system through film badges or other personal radiation recording devices so that the actual radiation exposure of every employee is known and any doses beyond established limits can be quickly spotted.

With these precautions and careful monitoring, the workplace where radiation is involved has a relatively low hazard level. When measured on the basis of known demonstrable ill effects on workers' health, the industries working with radiation have actually been among the safest we have. If you work with radiation, however, don't be afraid to insist on adequate protection and monitoring to make sure your radiation dose is "as low as reasonably achievable."

AIR TRAVEL.. While many people are concerned over the safety of air travel, they hardly every include radiation as one of the risks to be considered. There is, however, a significant increase in levels of natural cosmic radiation with altitude. At the levels of commercial jet flight, about 30,000 feet, a 2-hour flight will expose a passenger to about 0.4 mrem of mixed cosmic radiation above the normal background received on the ground at sea level. At supersonic flight altitudes of 60,000 to 70,000 feet, as typically reached by the Concorde, for example, the radiation level will be double again. This is a relatively small increase for a passenger, even one taking 10 or more flights per year. But for the pilots and crew who typically fly 720 hours a year, the cumulative dose at normal jet altitudes will be about 160 mrem/yr. Their annual total dose of radiation from natural sources would thus be 260 mrem, 100 from normal ground level irradiation and 160 additional from flying. Military pilots and the crew of the Concorde might total as much as 400 mrem/yr or about 4 times normal.

There has as yet been no connection made between these amounts of radiation and any harmful health effect to the pilots or crews. This is consistent with other information concerning the effects of such very low levels of added irradiation. A good example is the data from Guangdon Province in China in which no measurable impact was seen from levels such as this even though some of the radiation was from material breathed or eaten and hence acting internally. So if you are going to worry while flying, worry about accidents and not about radiation.

MICROWAVES. We now return to an area in which there is still a great deal of controversy and where the risk/reward issues are more complex. Basically this is because the risk part of the equation is, as we have seen, less than crystal clear. If there really is no physical harm from microwave intensities below 10 mW/cm² as the U.S. standards imply, then the presence of high-power microwave antennas near high-rise apartment complexes in our large cities should not be of concern to us from a health standpoint. We have seen, however, that scientific controversy still exists as to whether this is a safe standard or whether there are subtle psychological and biochemical effects existing at levels lower than 10 mW/cm². The public is on the one hand assured by groups such as the NCRP that adequate standards and controls exist and on the other hand sees books such as Paul Brodeur's *The Zapping of America* published in 1977 with the cover-jacket statement: "Microwave radiation can blind you, alter your behavior, cause genetic damage, and even kill you. The risks have been hidden from you by the Pentagon, the State Department and the electronics industry. With this book the microwave cover up is ended."

It is unfortunate that this kind of sensationalism is permitted to becloud a complex and important issue. Yes, it is true that microwaves can blind you, alter your behavior, cause genetic damage, and even kill you if they are focused on you in sufficient intensity for a long enough time. Yes, it is true that in the Moscow Mystery many elements of coverup existed between the United States and the USSR, between the State Department and its employees, and between the State Department and the public. A more open and direct approach would have been of benefit to all. It is not true, however, as the publisher's description implies, that a sinister conspiracy exists or existed between the military, the State Department, and the electronic industry to hoodwink the public about the hazards of microwaves.

Science doesn't work that way, but science has its biases and they frequently show. The U.S. position on microwave standards reflects a scientific bias originating in the nature of the early U.S. work and the assumptions underlying it, particularly the assumption that the only significant effects from microwaves were those arising from the temperature increases the microwaves induced. When challenged, science can be persuaded to examine its biases. This is what is currently underway in research on the effects of microwaves.

But how can the public achieve a rational perspective to govern its actions towards microwave installations and other related issues? Its greatest salvation is common sense, and that is what we will try to apply to this issue now. The application of common sense must first consider the motivations and driving pressures on each side in the debate. This provides a perspective on their positions and illuminates the realities of the issue. It is certainly true that one of the key objectives of our military services is to provide security against enemy attack. Radar installations are vital to this objective, and in general the more powerful they are the more effective they are. The safety of operating personnel is another important objective, but risks to military personnel are inherent in their calling and to a limited degree are acceptable to achieve the desired overall security. It is thus reasonable that in the absence of strong evidence of harmful effects the military services will push microwave power levels as high as possible and will seek experimental evidence to justify these levels. Once a set of standards has been established, equipment will be designed so as to just meet these standards, because that is usually the most economic way to achieve the military objective. Anything threatening to lower these standards can carry severe economic penalties and will be vigorously and sometimes obstinately contested.

The military services are not generally populated by knaves and fools, even though the occasional knave or fool who does exist there (as in any large organization) sometimes makes us forget this. There is a basic recognition of the need to operate safely and the cost in money and prestige of not doing so. Thus when challenged the services can mount effective research programs to establish what the real facts are. That the research work itself must draw on scientists or doctors from independent organizations such as universities, hospitals, companies, or nonprofit research centers is further assurance that the work will try to find the truth rather than obscure it or cover it up.

In similar vein, the operators of TV transmitters or the manufacturers of microwave ovens desire to minimize operating or manufacturing cost. They also have a powerful incentive to ensure that they are not harming the public. While a moral incentive is involved, it is only realistic to recognize that there is an overriding economic incentive as well. Harming a member of the public and particularly your customer these days can be a large step towards bankruptcy. Commercial research and development, however, sometimes has a greater tendency towards defense of the status quo than an objective seeking of the truth, and we have to be aware of this possibility in evaluating commercially supported research. The work supported by the tobacco industry attempting to find valid reasons for continuing the habit of smoking is a case in point. This is not to imply in any way that all commercially supported work in product safety is suspect. Most of it is honest and valid, but common sense requires that individual cases at least be scrutinized.

The official and unofficial protectors of our environment see the issues from a different perspective. They can afford to take the position that the only concern is absolute safety. If that poses a problem to the generator of the technology, he can just find a way around it. Either improve the technology until it is absolutely safe or go out of business. The emphasis may therefore be very much on the risk and very little on the reward. Research supported by government environmental protection agencies or by independent research groups supported by or motivated by environmental concerns may be biased towards finding effects at lower and lower levels of the suspected harmful agent, whether it be microwaves or toxic wastes.

It is probably fortunate these differing viewpoints exist because the debate they generate inevitably leads to compromise that considers the economic cost of standards in balance with realistic risk considerations. This is the process now underway in the review of microwave standards. The extreme and unusual difference in viewpoint between scientists in the United States and the USSR has made the process more complicated than normal and leaves the public with an undesirable degree of uncertainty. Be assured, however, there is nothing in the data from either side that indicates continued adherence to current U.S. standards leads to any likelihood of permanent or life-threatening harm. The uncertainty lies in not knowing whether temporary but annoying and significant effects may be occurring that have microwaves with intensities of 10 to 1000 μW/cm^2 as their unrecognized source. We have seen what the evidence is that these effects may exist. Now we need to see whether any of us are being unknowingly exposed to these levels of microwaves.

You will recall that the major sources of microwave radiation are military radar installations, microwave relay stations, satellite communication sites, aviation radar, police radar, and TV and FM broadcasting stations. A 1976 study by the U.S. Environmental Protection Agency (EPA) analyzed the relative contributions of these various sources to the possibility of incidental exposure of the general public to microwave intensities above 10 μW/cm^2. They concluded that while satellite systems and military radar have extremely powerful beams, they are directed away from populated localities and have adequate exclusion zones to keep irradiation levels to the general public under 10 μW/cm^2 during normal operation. Lack of care in maintenance, for example the presence of undetected mechanical or electronic misalignment of the equipment, could presumably cause problems, but regulations require continuous perimeter measurements of microwave levels to detect any such aberrations and permit their correction. If you live in the near vicinity of any such major installations, however, you should be sure that adequate monitoring is carried out both by the installations and by independent government agencies. With these precautions the risks are acceptably small.

VHF-TV and FM transmitters have the characteristic that their purpose is to reach your home. In this way they are fundamentally different from the radar and satellite communication systems that are designed to avoid homes and people. To be effective the signal reaching your TV antenna need not be strong, since the electronic systems for detecting and amplifying the incoming signals are far more sensitive than your untuned body. Nevertheless to reach outlying regions the signal strength near the transmitting antennas must be relatively strong, and EPA measurements have shown that nearby FM and TV transmitters are the main source of exposure to levels of microwave radiation above 10 μW/cm^2. Between 1976 and 1978 this agency carried out measurements at representative locations in 12 U.S. cities. They found that half the population of these cities (totalling over 38 million people) received microwave radiation levels less than 0.005 μW/cm^2 and half received levels above this. Less than 1% were exposed at levels above 1 μW/cm^2. But some areas were noted, typically in the upper stories of buildings near major TV or FM transmitters, where measurements were slightly over 100 μW/cm^2. These were the highest levels recorded. While a factor of 100 lower than the U.S. standard of 10 mW/cm^2, they are clearly in the region where small temporary psychological and biochemical effects have been observed in animals. They have not caused proven harm in humans, but prudence suggests these levels should be avoided until the experimental evidence on their effect or the lack thereof is clear and unequivocal.

Thus anyone living within one-quarter mile of a major TV or FM broadcast antenna would be well advised to obtain actual measurement of microwave irradiation levels in his or her apartment or home. If measurements are recorded significantly above 10 μW/cm^2, proper screening should be installed to reduce the level at least below 10 μW/cm^2. This is advisable because residence in such a home would mean continued exposure over long periods of time. The occasional visit to the top of the Empire State Building where microwave levels of 30 μW/cm^2 have been observed poses no problem, however. The city traffic noise will probably jangle your nerves much more than the microwaves.

It is important to understand this point about continuous versus occasional encounter with low levels of microwave radiation because the uses for microwaves are continuing to expand and we encounter them in more places every year. The police use radar to check our speed on the highway. Pleasure yachts as well as commercial vessels have turned increasingly to radar as a navigation aid. Private microwave communication systems are increasing, and the possibility exists for extensive use in automobile telephone systems. Microwave systems are being installed at the exits of stores to catch shoplifters through the recognition of signals reflected from special tags on the merchandise. These tags are removed or covered when a legitimate purchaser pays at the checkout counter. All of these uses are in someone else's control, not ours; yet we receive, usually unknowingly, some of the radiation involved. The levels of irradiation will be very low, however. Police radar, for example, has intensities less than 1 μW/cm^2 at 44 feet from the source in the police vehicle. The radiation we receive is therefore both very weak and intermittent. Each other source is, by itself, similarly harmless.

We need to be able to feel comfortable that these increasing uses cannot somehow accumulate until they really do impact upon our health. Many factors combine to develop this feeling of comfort if we will use our common sense in considering them. First and foremost, the radiation levels and times are below even the most conservative suggested standards. Second, governmental regulatory and legislative bodies are well aware of the possible problems and continue to act as our watchdogs. Our third bulwark is the network of scientific organizations that monitor the area and carry out research on all aspects of microwave effects. Finally and very importantly we have an additional network of organizations dedicated to the protection of our environment. They can be counted on to raise the alarm if industry or the military for any reason becomes insensitive to our safety. Microwaves can be expected to be increasingly useful to us in many ways. They can be expected to achieve their benefits at zero risk to our health if we continue to respect them and use them wisely.

HIGH-VOLTAGE TRANSMISSION LINES. A recent television drama portrayed the fight of a farmer against the dictatorial electric utility, which wished to obtain the rights to erect a high-voltage transmission line across part of his farm. To fight this dastardly invasion the farmer mobilized the good people of his community to take vigilante action against the utility and its president. The fight was won when his cohorts in one night were able to erect a full-size transmission tower several hundred feet high precisely in the center of the utility executive's front lawn. This made the utility executive realize how the farmer felt, and the power line was cancelled.

This ridiculous simplistic story shows one side of the issue of high-voltage transmission lines. Seen at close range by the owner whose farm they cross, the transmission lines are ugly mainifestations of man's capacity to defile natural beauty. Yet our cities cannot exist without power. Power transmission from Niagara Falls, for example, to the cities of New York and New England is absolutely essential to the well-being of these regions. Our system of laws recognizes this in its provisions for the establishment of rights-of-way for power lines. It recognizes that the benefits to many must at times take precedence over the loss to a few. It also implies clearly that the loss should be kept as low as possible and that it be compensated for fairly. One can sympathise with the anger of the farmer whose land has been invaded, but at the same time one must recognize the importance of the need to transport the power.

In considering the environmental impact of a high-voltage transmission line, the effect on the view is only one factor. The electric and magnetic fields associated with the line can cause annoying static in radio and TV reception, although this can be reduced by electronic filters. There is often a sizzling noise in damp weather caused by minor electrical discharges into the damp air surrounding the power lines. This is the phenomenon known as corona. Also large metal objects such as a car or tractor very close to the line can become electrically charged in such a way that an individual touching them can receive a spark discharge shock. This is the same phenomenon as the static discharge shocks we encounter at home when we walk on certain kinds of rugs on a dry winter day and then touch a metal object or even another person. These shocks encountered beneath a high-voltage line will be considerably larger but are not large enough to cause permanent injury.

These are all annoyance factors, however, and as such can be compensated for by negotiated cash settlements. Human nature being what it is, we cannot expect these negotiations always to be amicable nor both sides always to be satisfied. It is therefore the role of the courts to make final decisions as necessary.

If a health hazard is presumed to be involved because of the presence of a new transmission line, the problem becomes more difficult. The creation of a nuisance on a man's property in order to greatly benefit society as a whole can be handled by trying to establish a reasonable monetary compensation for the nuisance. A threat to a man's health on his own property is much more serious. While even that could be settled by abandonment of the property and full value or more than full value reimbursement, this is a much more emotional issue. It hits at the deeply ingrained concept that "a man's home is his castle." The health issue is often used in attempts to block the construction of vital power lines. Utilities must get approval from public utility commissions for siting of power lines, and this approval usually involves public hearings. "Environmental impact" statements may be required, and there may be public hearings also.

Individuals wishing to block the power lines can use these hearings to get legitimate questions discussed and answered, but they can also use them as tactical battlegrounds in which the intent is not to get questions answered but to delay, obstruct, and prevent or require rerouting of the power line. The question of health effects can thus become a political rather than a technical issue with both sides in the dispute trying to muster scientific opinion favorable to their postions.

The public is caught in the cross fire and may find it difficult to locate the truth in the haze of conflicting claims. In this case, some unemotional recognition of essential facts leads to a relatively clear conclusion. First, it is quite clear that the electrical and magnetic fields associated with all high-voltage transmission lines up to 765 kV are not high enough to cause permanent health effects even to someone spending all of his time standing below the lines. Second, it is also clear that temporary effects caused by any normal degree of presence of man or animal in the field are minor and should not be a factor in consideration of whether or not the line should be built. Finally, there is no rational alternate way of transporting the enormous amounts of essential power carried by these lines. Underground cables are far too costly, microwaves are too hazardous and are not capable of transmitting adequate amounts of power. The idea of on-site generation of power by fuel cells, solar systems, or windmills is simply not practical for large population centers. We therefore have a highly safe but annoyingly evident means of power transmission that is absolutely essential and for which there is no reasonable substitute. Our job is to see that the lines are built when needed, but built on intelligently planned and fairly compensated rights-of-way.

PROJECT SANGUINE. The people of the states of Michigan and Wisconsin have had to consider their reaction to a unique proposal to add another form of radiation to their environment. This is the story of a project conceived in 1958 as Project Seafarer, later renamed Project Sanguine, and recently revived as Project ELF (for Extemely Low Frequency). The driving force for this concept was the need to communicate to our fleet of nuclear submarines beneath the oceans in all parts of the world. Standard radio or microwave communication frequencies cannot penetrate appreciably beneath the ocean surface and are useless in trying to reach a deeply submerged submarine. It was recognized that extremely long wavelength waves, similar to those emanating from 60 cycle (60 Hz) electric power transmission lines, would penetrate both earth and ocean and could be modified or "modulated" to carry messages just as in conventional FM radio transmission.

A base frequency of 76 Hz was selected as being best suited for the task. This signal has a wavelength of 2500 miles, however, and for efficient transmission the antenna sending out the signal has to be very long also. The proposal envisaged a transmitting system with 47 antennas, each from 19 to 96 miles long, laid out in a grid pattern with 3.7 miles between parallel antennas buried from 2.5 to 6 feet in the ground. The total system required an area of approximately 4000 square miles in the 1975 plan for a site in northern Michigan.

For a properly radiating electromagnetic signal the electric current (about 100 amperes) had to flow from one end of an antenna segment to the other and then pass into the ground and flow back through the ground to the originating end of the antenna. Thus a great loop would be formed carrying electric current and generating the electromagnetic waves that could carry across the country and down through the ocean to the listening submarines with suitable receiving systems. Because of the particular conducting characteristics in the underlying bedrock, only a few areas in the United States were felt to be suitable for such an antenna system. They included the upper peninsulas of Michigan and Wisconsin, south central Texas, and parts of Nevada and New Mexico. Most serious attention was focused on Michigan and Wisconsin, however.

Research on the possibility of ELF transmission had begun in the late 1950s, but the general public first really became aware of it in 1968 when a Wisconsin congressman announced that Project Sanguine would be located in a large area of northern Wisconsin, including his district. Public curiosity was soon followed by public objection based on fear of the effects of the electromagnetic radiation or the associated electric fields or both. The bitter debate ensuing resulted in deferral and reevaluation of the project, and in 1976 a detailed study of the possible radiation effects was undertaken by a committee of the National Academy of Sciences. Their report was published in 1977 and contained the following conclusion:

The Committee's considered opinion is that such fields will not cause a significant and adverse biologic disturbance, except in the event of electric shock, which is of serious concern. In fact, apart from the possible result of electric shock, the Committee cannot identify with certainty any specific biologic effects that will definitely result from exposure to the proposed Seafarer fields.

We need to look into the evidence behind this conclusion to understand it and accept it. As designed in 1975 the entire system of 47 antennas would draw 14 MW continuously. Both electric and magnetic fields would be generated and we will consider the electric field and its effects first. It could be readily calculated that the electric field at ground level directly over a buried cable would be 0.07 volts per meter (0.07 V/m). A meter is close to a yard in length and is about the length of an average man's stride. Thus a man walking at this point would have a voltage difference of 0.07 V between his two feet. A single flashlight cell has a voltage of 1.5 V or 20 times as much as this. You should recall also that electric fields *in air* under high-voltage transmission lines are as high as 5 to 10 kV/m or about 1000 times as high as Seafarer fields. These fields are in air, however, and in the Seafarer case the voltage difference is in the ground where much better coupling to a walking person is possible.

In considering the effects of electric fields the critical concern is not the voltage but how much current the voltage can cause to flow in the contacting person's body. A current of 1 milliamp (1 mA) can just be sensed but causes no harm. However, 10 mA can cause muscle paralysis; for example, a person holding two electrical contacts producing a 10 mA flow in his arms cannot let go of the contacts. At total body flows of 40 to 100 mA the heart will go into a rapid uncontrolled flutter known as fibrillation, and death can result. The Underwriters Laboratory limit for human safety is specified as 5 mA maximum.

The important point is to ensure that body currents are kept below 5 mA and preferably below 1 mA. But how does this relate to voltage? The classic law connecting voltage and current in direct current (DC) flow states that current (amps) is equal to the voltage (volts) divided by the resistance (ohms) or $I = E/R$. Thus the higher the voltage, the higher the current if the resistance stays constant. We must remember that for Seafarer we are talking about alternating current (AC) rather than DC. While the relationships are somewhat more complex, the same general principles hold, however. In either case there are two resistances that have to be considered. First is the intrinsic resistance of the body itself (about 1000 ohms), and second is the contact resistance. Most of you have had to make electrical connections and know that they work best when the wires and contacts are cleaned with a knife or similar tool to ensure a low-resistance contact. Similarly if a person is standing in bare feet in a puddle of water where a ground voltage exists as above a Seafarer grid, there will be good electrical contact from the ground to the person. If the individual is on dry ground in a pair of rubber-soled shoes the contact resistance will be very high and little current can flow through the person. But in determining system safety we have to consider the worst possible situation, namely, one where contact resistance is essentially zero.

Let's consider the case of two men carrying a 15-foot aluminum canoe, one man at each end, and assume that the men are barefoot in wet weather. 15 feet is about 5 meters, and at 0.07 V/m there will be 0.35 volts between the two men. Their combined resistance will be 2000 ohms, assuming 1000 ohms for each man and zero resistance for the canoe. In an oversimplified but still reasonably approximate view the current through the men would be 0.35/2000 or 0.175 mA. This is well below the 5 mA threshold of concern and indicates that even in this extreme case the electric fields above the antenna grid would not pose a hazard. Thus work or play in the area above the antenna grid would be perfectly safe even if long conducting objects were to be handled there.

The National Academy of Sciences report pointed out, however, that in the areas at the ends of the antennas where the electric currents went into the ground, the electric fields would be higher than in the ground above the antenna grid. In these grounding areas the fields could reach as high as 15 V/m. Fields of this magnitude could, under conditions where long conducting objects such as canoes or tractors with plows were involved, create the possibility of electric shock conditions that could be harmful. Recommendations were made to change the designs of the ground terminals so as to eliminate this possibility.

No other significant effects of the electric fields associated with Seafarer were found. Topics considered were genetic effects, fertility and growth effects, physiology and biochemistry, cell growth and division, sleep-wakefulness cycles, behavioral effects, and others. The effects of the magnetic fields associated with Seafarer were also studied. These magnetic fields could be as high as 0.11 G, but the normal earth magnetic field is 0.5 G and the fields associated with power tools and electric appliances can be as high as 1 to 10 G and have no known health hazards.

The possible effects of both types of fields on mammals, fish, birds, bees, other insects, and even bacteria were also considered, and no harmful effects could be identified. Some reservations were expressed concerning possible interference with the navigational skills of bees or birds. In the latter case it is known that some birds are sensitive to steady magnetic fields and may even take advantage of naturally existing ones in their extraordinary navigational feats. The Seafarer fields are alternating ones, however, and logically should not upset these systems. Nevertheless we must admit here an area of ignorance.

In spite of setbacks in the late 1970s the Seafarer Project is not dead. The navy has found no good alternative to this long wave approach to communication with submarines, and the concept has revived as Project ELF. In its present form it is to use 56 miles of new antenna on state-owned land in the Upper Michigan peninsula and will also utilize 28 miles of existing antenna at the Clam Lake facility in Wisconsin, which had previously served as a test site. In July 1983 the Michigan Natural Resources Committee granted rights to the navy permitting installation of the new antenna lines. This was in spite of objections from Michigan's governor and one of its senators. The governor of Wisconsin requested an injunction to stop work on the project, and a long battle is expected before ELF is either completed or abandoned.

It is a classic issue requiring individual judgements as to the balance of risks and rewards. The people in Michigan and Wisconsin must accept whatever risks exist, but they will only very indirectly reap the rewards. Submarines beneath the Atlantic Ocean, even though part of the nation's defense bulwark, are not part of the everyday concerns of a Wisconsin farmer. The National Academy of Sciences report makes a convincing case that as far as local residents are concerned they will not even know the facility exists except for the sight of antenna wires (now planned to be above ground) and the disruption caused by their installation. With a high degree of certainty there are no health risks, yet this is not an absolute certainty. Some environmental impacts such as on the behavior of migratory birds are not fully understood, yet the probability of serious harm seems very small. Thus the overall risk seems tiny, and the critical focus should perhaps in this case be on the reward. How great is it? Can it be achieved in other ways? We often ask the question "Is it worth the price?" In the case of ELF the evidence is very strong that the price in health hazard is extremely small or nonexistent. Thus the final judgement concerning the desirability of ELF should focus on its dollar cost in relation to the national welfare, not on concerns about the environmental impact of its radiation. The communities affected, however, should be sure the project in its final design form incorporates changes that will reduce ground surface voltages to the lower levels suggested by the National Academy of Sciences study.

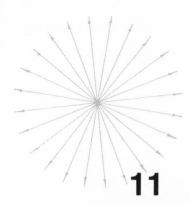

11

NUCLEAR ARSENALS

A NUCLEAR WEAPON CONTAINS A TRILOGY OF DEATH: blast, heat, and radiation. These are separable in a technical sense, but in a social and political sense it is futile to treat them separately. In this chapter we will consider briefly the whole issue of nuclear weapons and their place in our society.

To begin we can turn again to concepts of reward and risk to help frame and focus our thoughts. One of the great goals of all thinking people in the world is peace. This was why the bomb was developed in the first place; those striving to achieve it saw it as bringing a quick end to World War II and a basis for a new and lasting peace. Their first goal, a quick end to the war, was dramatically attained. The second goal, a lasting peace, at least between the world's great powers, has survived for over 40 years but seems far from assured in the future. The continuation of peace is based on the power of mutual threat, a power both sides have continually seen fit to escalate lest the other antagonist gain a decisive advantage. The proliferation of nuclear weapons by the United States has been based primarily on two factors: (1) that their presence must act as a deterrent to any major acts of aggression by the enemy and (2) that by having more and better weapons the United States could emerge from a nuclear war with its ideals of freedom still intact and with a world still reconstructable in the old image.

We must first analyze the reality of these rewards. Are they credible and are they attainable? The deterrence argument is at least partially valid. It has worked so far, since no nuclear weapon has been used in anger since Nagasaki. The critical question, however, is whether it can continue to work. A typical estimate of the size of the U.S. nuclear arsenal puts it at 7000 to 9000 strategic warheads and 17,000 tactical nuclear weapons. A "strategic warhead" is presumably a weapon anywhere from 20 to 1000 times as powerful as the weapons used at Hiroshima and Nagasaki. A "tactical nuclear weapon" could range from sizes well below to somewhat above the approximately 20 kiloton weapons used on Japan. The destructive potential of these weapons is beyond our power of imagining. One need only look at the pictures of the totally devastated wasteland that was Hiroshima after the bomb and try to imagine a scene in which the great cities of the United States or the USSR or Europe are similarly leveled, the great hospitals are gone, communication systems destroyed, water supplies contaminated, and death and sickness stalk the surviving fringes. Yes, Hiroshima and Nagasaki live again today, but the bombs that struck them were puny in contrast to those available now. Beyond their city limits was an untouched society ready to mobilize to come to their aid and strong enough when peace came to rebuild those cities so that now they are finer than before. The technological world today, however, depends on an intricate system of communication, transport, sanitation, and the control of law. As more and more cities were destroyed this system would disintegrate leaving humanity only its instincts and personal survival skills on which to continue existence and few resources with which to rebuild.

This picture, readily comprehensible to the leaders of all great nations, is certainly a powerful deterrent to the initiation of nuclear war. The threat of instant and unpreventable retaliation in kind makes even the most aggressive or paranoid head of state pause before starting an attack.

Deterrence is therefore real, but unfortunately it is not absolute. A false electronic signal indicating an oncoming enemy missile could trigger premature retaliation and precipitate the war. The use of a single weapon against a nonnuclear country to punish it for aggressive behavior might escalate through retaliation by its nuclear-armed allies. The delusions of a modern Hitler might start with tactical weapons to aid invasion and slowly lead to total and final nuclear involvement. The reward of deterrence, therefore, is an uncertain one.

In the years since Hiroshima and Nagasaki the two great protagonists, the United States and the USSR, have continued to increase the number, the size, and the variety of their nuclear arsenals. In part this has been because of their belief that an even balance of power is the best deterrent coupled to the unwillingness of either side to recognize for any appreciable time that such a balance actually exists and only needs to be maintained at the status quo. Implicit in the actions of the leaders of each side has been the belief that even if war came, as long as they could maintain a defensive or offensive edge, they could finally win and preserve their own national goals and ideology. This was the second potential reward of nuclear weapons, and of the two hoped-for rewards it is the most treacherous. It is treacherous because it is an almost instinctive belief, and yet because of the power of devastation of nuclear weapons it is unlikely to be true.

We in the United States use freedom as the watchword of our society. We treasure the rights of individuals to free expression, freedom of movement, freedom of individual choice in the way we live. When we analyze our world, however, we find that these freedoms are not absolute, and our real goal is the preservation of those freedoms to which we have become accustomed and which we have elevated above other freedoms. We accept that we are not free to kill, to steal, or to physically abuse our neighbors. We somewhat grudgingly give up the freedom to travel as fast as we would like on the superhighways. Suburbanites acquiesce in the restrictions against building too close to the lot line and thus offending the neighbor next door. Young men and women, by law at least, give up the freedom to drink alcoholic beverages until they are 18 or 20 or 21 as their state law dictates. These restrictions have been determined to be valid limitations of individual freedom that benefit the society as a whole. We cherish, however, the right to speak our mind about the problems and actions of government, the right to choose our leaders, the right to privacy in our own homes, the right to travel freely when and where we choose.

The "freedom" that we fight and die for is really the set of freedoms and restrictions that make up our current way of life contrasted to the set of freedoms and restrictions of alternate ways of living such as socialistic or communistic or totalitarian systems. The preservation of our present pattern of freedoms and restrictions during a nuclear war and its aftermath would be impossible. The imposition of martial law in devastated cities would be essential. One has only to remember the looting and rampaging that occurred in New York City in the electrical blackout in 1977 to recognize how thin is the veneer of our civilized and orderly behavior. Conscription, censorship, and information control, as well as restrictions on movement to and from disaster areas would be the order of the day. Whether we could ever return to a semblance of our former social structure would depend on the extent of moral and ethical changes the enormity of destruction would cause in the hearts and minds of the survivors. Those changed and bitter survivors would have to piece together a new society with whatever outside help they could find. We cannot predict what form that new society would take. We can only recognize the incredible cost in human life and physical destruction paid to try to preserve the old society, which will have vanished as surely as if we had been overrun by our enemy without a fight.

Be assured that there would be survivors. There are physical limits to the destructive range of the blast and heat effects of even the largest weapons. We pointed out earlier that a 100 megaton weapon, the largest ever considered, would destroy buildings up to 30 to 35 miles from the explosion point and start fires up to 70 miles away. But even multiple weapons explosions in large city areas would leave populated regions 100 or 200 hundred miles away relatively undamaged. Fallout radiation would cause many deaths downwind from the explosion areas, but populations upwind would be spared even this hazard. A major nuclear war in the near future would probably involve the USSR and the United States where casualties could be enormous. Yet Africa, South America, and perhaps China and the Far East might remain relatively untouched, so humanity would continue to exist.

But *our* question is whether the United States and our allies, the so-called society of the Western world, could continue to exist. That society and what we believe is its ideal balance of freedom and control is what we are trying to preserve. That is why we have nuclear weapons. And that is what would inevitably disappear.

Our line of reasoning has thus led us to see that deterrence through the mere presence of nuclear weapons is uncertain at best and cannot last forever. It has also shown that if deterrence fails, as it probably will, the increase in size and variety of nuclear weapons can only lead to increased levels of mutual destruction and increased certainty of the elimination of the very societal benefits the weapons were supposed to protect. The inescapable conclusion is that nuclear weapons cannot achieve the rewards for which they have been created and that the risks they pose to our society are intolerable. Our every effort must be directed towards their reduction in size and number and their eventual elimination. How this can be accomplished is beyond the scope of this book, but our long-term survival as a civilized country depends on its being accomplished.

The subject of this book is radiation, however, and we cannot leave the topic of nuclear weapons without further discussion of that aspect of their trilogy of death. There has been much discussion of civil defense against nuclear weapons. This has included fallout shelters, which could act as partial protection against heat and blast and in particular isolate against radioactive fallout. Such fallout shelters should have effective air filtration systems and be well stocked with food and water for long occupancy while outside radiation decays. There has been little movement in this country towards universal construction of such shelters partly because of apathy and partly because of the large cost involved. Recently the trend has been towards plans whereby the population of large cities would be evacuated to the suburbs to provide safety through dispersal. These plans assume some advance warning of nuclear attack, and many critics feel that their practical application would be nearly impossible.

It is not surprising that most individuals in the United States today simply ignore the whole question of the nuclear weapons threat. But the threat is too real to be ignored. At the very least every man, woman, and child in the United States should learn the elementary principles of the effects of nuclear weapons. Knowledge of these principles could be the basis for survival rather than death in the fringe region where survival is possible away from the central point of the explosion. In particular, understanding of the behavior of fallout radiation will be critically important. Cleanup and disposal of radioactive dust and debris, adequate shielding, evaluation of the probable safety of food and water, the timing and direction of flight to an uncontaminated area will all depend on knowing as much as possible about the characteristics of radioactive fallout. Two fundamental reference books are *The Effects of Nuclear War* prepared by the Congressional Office of Technology Assessment and available from the Government Printing Office, Washington, D.C. 20510 and *The Effects of Nuclear Weapons* by Samuel Glasstone published in 1977. Glasstone's book is by far the most specific and useful. Reading it and having it available is probably the most cost-effective action that you can take to improve your chances for survival in a nuclear war. The next most useful act would be to obtain a good general purpose radiation detection instrument and learn how to use it. Like the flashlight in the drawer, which is essential for use in a nighttime power shortage, a radiation detection instrument can be your salvation if nuclear war ever comes. The best line of defense, however, is preventing such a war in the first place. The only sure path to prevention is the total elimination of nuclear weapons from the arsenals of the world. If it takes unity of purpose, we must provide it. If it takes leadership, we must lead. If it takes compromise, we must compromise. But we must do it and do it soon.

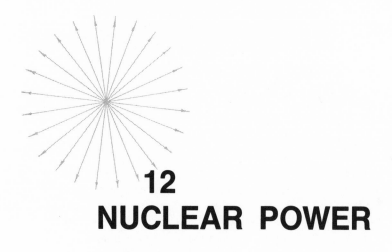

12
NUCLEAR POWER

THE EXISTING OPPOSITION TO NUCLEAR POWER stems from three areas: (1) concern about spread of lethal radioactivity from a reactor in case of a major accident, (2) concern about the ability to contain safely over long periods of time the high level radioactive wastes generated, and (3) concern that the existence of nuclear power will make the spread of nuclear weapons more likely. We need to analyze each of these concerns in more detail before grappling with the thorny problem of whether the rewards of nuclear power more than balance the scales.

RISKS OF A NUCLEAR ACCIDENT

Rasmussen estimated that the very worst possible accident with a reactor of today's design might kill 3300 people outright and cause 1500 additional cancer deaths annually in the period 10 to 40 years after the accident. This would be a total of 46,500 deaths traceable to the accident. The probability is that such an accident would only occur once in a million years in a world with 1000 operating reactors and would result in fewer deaths than are caused by automobile accidents *every year.*

There is a greater possibility of lesser accidents with lesser consequences, but we cannot define these risks very precisely. We can only point to the 29-year history of commercial nuclear power to date with no record of accidents with any direct consequences at all to human health. The Three Mile Island accident, to be sure, caused enormous inconvenience and considerable psychological shock to residents in the surrounding communities. There was no physical harm to anyone, however.

You may properly ask why you should believe these estimates of the consequences of the worst possible nuclear accident. Many groups opposed to nuclear power make far more spectacular predictions of the frequency with which such accidents will occur and their severity when they do. As in any search for reality you can only base your perceptions and beliefs in what has actually happened to date and on reasoned analysis of the basis for predictions in the future. The Rasmussen study is the best approach to reasoned analysis of the future. While not gospel, it was enormously detailed, used the best methodology available, and used it with a high degree of objectivity. For those reasons it is probably as good a guide as any and should not be tossed aside lightly by either pronuclear or antinuclear sides.

Even the Three Mile Island accident, which involved major damage to the reactor core, is still within the probabilities established by Rasmussen. This accident has been cited by pronuclear forces as a proof that even a very serious malfunction coupled with equally serious operator errors still led to no physical harm to the surrounding community. Thus it proves that reactors are not only foolproof, but damn-foolproof as well. The antinuclear forces point out that the accident was very close to much more serious consequences and that these will inevitably occur much more often than Rasmussen predicts. Which of these positions is nearer the truth? No one can say with certainty, but there are a number of factors favoring the long-term safety of nuclear reactors. A malfunction threatening the health or lives of residents outside the reactor must be accompanied by a major breach of the containment building. This has never happened. Small leaks yes, but no major rupture. Postaccident analysis of the Three Mile Island episode shows that it was not really close to happening there. The wild speculations at the time of the accident were just that: wild speculations based on insufficient knowledge of what was really happening. This does not mean that containment breach is impossible. It can happen, but it is very unlikely. It is less likely now than it was at the time of Three Mile Island because the industry learned some costly lessons from that episode and has significantly improved its safety practices. A Nuclear Safety Analysis Center (NSAC) has been established by the electric power companies through the Electric Power Research Institute (EPRI) to assist electric utilities in the safe operation of nuclear plants. The Institute of Nuclear Power Operations (INPO) has also been formed by EPRI to carry out regular reviews of the operating procedures of all nuclear power plants and ensure their adherence to good safety standards. The Nuclear Regulatory Commission (NRC) has continued to sharpen its regulations aimed at safe reactor operations. These steps will not make the safety record perfect, but they will certainly improve it.

There is another new and very important consideration that has entered the picture since Three Mile Island. It involves the factor known as the "source term." Simply stated, the source term is the assumption in a nuclear plant risk analysis estimating the percent of major fission products that will escape in a given type of accident. The earliest estimates of these quantities were made by Brookhaven National Laboratories in 1957 and were knowingly set on the high side to ensure conservative treatment of possible risks. They assumed that in the maxium credible accident 50% of the noble gases (xenon and krypton), 50% of the iodine and related chemicals, and 50% of the solid fission products would escape outside the containment. It was soon recognized that the estimate of 50% of the solid fission products escaping was much too high, and the estimate of only 50% of the noble gases was probably low. Thus a later Atomic Energy Commission study revised the source terms to be 100% of the noble gases, 25% of the iodine, and only 1% of the solid fission products. The Rasmussen report in somewhat more analytical fashion varied the source term with the type of accident, but for their worst-case accident assumed iodine releases to the atmosphere of 40 to 50%. These isotopes are prominent in the fatality calculations.

In the Three Mile Island accident, even though the reactor core was substantially damaged, only about 18 curies of radioactive iodine were actually released into the atmosphere out of 28 *million* curies contained in the fuel. This is only 0.00008%. The bulk of the iodine radioactivity was contained in the primary water-cooling system or deposited within the containment building. It became clear from the locations of the iodine radioactivity and the way it was deposited that even if there had been a major rupture of the containment building, only relatively small amounts of the radioactive iodine and cesium would have been released into the atomosphere. This indicates that the source term used in the Rasmussen calculations and most other estimates of reactor accident tolls may have been far too large, and thus their predictions of fatalities and illness were far too high. In other words the hazard from even the most severe and unlikely reactor accidents may be much less than we have predicted.

While this evidence has been hailed by nuclear proponents as further justification for nuclear power, it has been openly questioned by those opposed. It is a very important factor, however, which must be carefully analyzed for verification or disproof. Several major studies have been undertaken recently to examine the issue. In June 1982 the American Nuclear Society established a Special Committee on Source Terms with membership from the United States, the United Kingdom, France, the Federal Republic of Germany, the Commission of European Communities, and Japan. Their report, issued in September 1984, concluded: "In general an ample foundation has been provided to warrant reductions of the source term estimates in WASH-1400 (the Rasmussen report) by more than an order of magnitude to as much as several orders of magnitude." A second industry study, the so-called Industry Degraded Core Rulemaking (IDCOR) study, sponsored by 63 nuclear utilities, architect-engineers, and reactor manufacturers, plus organizations in Japan, Thailand, Sweden, and Finland, provided the NRC a detailed 48-volume report in November 1984 concluding that a severe nuclear plant accident would release only about 10% as much radiation as previously calculated.

The NRC itself sponsored another study by the American Physical Society in a move to avoid industry biases. While agreeing with the general thrust of the other two studies, the American Physical Society study cautioned that different reactor types could behave in different ways and that still more thorough analyses should be undertaken before relaxing the stringency of NRC safety regulations based on the Rasmussen estimates.

Even with proper discounts for any biases existing in the industry-sponsored studies, it seems almost certain that the radioactivity releases from any conceivable reactor accident will be significantly less than had been estimated 11 years ago (1975). The difference primarily is that the radioactive iodine is not released as gaseous iodine but combines, particularly with cesium, to form water soluble salts that stay in the reactor water or are deposited within the reactor containment system. Even should the containment be breached, the dispersion of the iodine would be far less than had it escaped as a gas. Thus it seems probable the casualty estimates for the worst possible nuclear reactor accident may be reduced to a few hundred lives or less, still with a probability of occurrence of only 1 chance in a million years with 1000 operating reactors.

It is undeniable that living near a nuclear power plant involves an element of risk of harm from a nuclear accident. It is clearly demonstrable that this risk is very, very low. Every home in every location has risks associated with it. All have risks from fire, varying with the type of home construction and the distance from the fire hydrant or other source of water. Homes near airports have risks from the crashes of errant aircraft. Rivers flood, dams burst, hurricanes and tornados occur, chemical plants have explosions spreading poisonous fumes, out-of-control automobiles crash into homes, and you can think of many more similar hazards. If these events are sufficiently unlikely, we accept them and don't think of them. Major nuclear plant accidents are in the class of possible but very rare events of which the consequences might be severe but are not different in magnitude from other much commoner disasters such as floods, explosions, or plane crashes. Are the risks worth the rewards? We'll come to that point later in this chapter.

DISPOSAL OF NUCLEAR WASTES

Common criticisms of nuclear power are that it continually adds to the levels of radiation in the world and that its wastes cannot be disposed of safely. The point is made again and again that high-level nuclear wastes remain radioactive for thousands of years and that no society should pass on to its descendants such a legacy of future hazard. This is an idealogical position with strong emotional appeal, but it needs to be analyzed with common sense and an accurate knowledge of the waste products and the characteristics of their radiation.

The fission process results in a myriad of fission products with enormously varied half-lives emitting α, β, and γ radiation. Additional elements heavier than uranium are formed also by neutron capture. These so-called transuranic elements typically have very long half-lives, up to thousands of years or more, and are usually α emitters. Spent nuclear fuel thus contains both high levels of fission products and the long-lived transuranics. In chemical reprocessing of fuels to recover fissionable plutonium and uranium, the transuranics are usually separated from the high-level fission product wastes. These latter products are categorized as high-level wastes, defined as irradiated fuel or its separated bulk fission products. Any wastes containing transuranic elements, such as plutonium, americium, or protactinium, for example, in amounts above 10 nanocuries (10 billionths of a curie) are considered transuranic or TRU wastes requiring separate labeling and disposal. This separation has only been required since 1970, however.

Fission products of very short half-lives such as minutes or hours, even though initially highly radioactive, are of little concern to us in problems of waste disposal because they will have decayed to insignificance in a few days or weeks. Very long half-life isotopes (a thousand years or more) are a problem because of their long term of existence, but this is leavened by their relatively low radioactivity per gram. If you have a pile of rocks and throw the rocks away slowly one at a time, the pile will last a long time. If you throw them away quickly in handfuls the pile will soon be gone. That is why long half-life isotopes are by nature low intensity emitters and hence less dangerous. They last a long time because particles are emitted less often.

The radioactive isotopes that are the greatest problem are those like strontium-90 with a half-life of 28.1 years or cesium-137 with a half-life of 30.2 years. They are initially strongly radioactive and, using a conservative rule of thumb of 20 half-lives to reduce their radiation close to background, they would be troublesome for 600 years. By that time, however, for every original curie there would be less than a millionth of a curie left.

Solidified high-level waste from all the estimated nuclear power production in the United States by the year 2000 could be fitted into a cube 50 feet on each side. The physical volume of these wastes is thus not very large, and the space required for its storage is not very large either. This presumes, however, that the wastes *can* be solidified and disposed of in that form.

Most of the high-level wastes today are either spent-fuel elements or high-level liquid wastes resulting from chemical reprocessing of the fuel. If the spent-fuel elements are stored without reprocessing, they will take up much more space than we have estimated above. But it still will not be an inordinately large amount. A typical reactor core is about 12 feet in diameter and about 12 feet long and contains from 200 to 800 fuel assemblies, depending on the reactor type. About one-third of these assemblies are removed and replaced each year, so even hundreds of reactors do not generate very large volumes of spent fuel elements.

Reprocessing to chemically separate the waste products from the uranium and plutonium permits the reuse of the remaining fissionable fuel material and also permits consolidation of the highly radioactive waste. Experimental work has proven the feasibility of converting the high-level liquid wastes into what are basically stable glasses. This is a process known as vitrification. It is in commercial operation at Marcoule, France, but has yet to be carried out on other than an experimental engineering scale in the United States. It is clear, however, that the process is feasible. Its implementation in the United States depends on basic decisions concerning the desirability of reprocessing and whether, if approved for commerical reactor wastes, it should be carried out by government or private industry.

We have shown by now that the volumes of high-level nuclear waste are not inordinately large but their hazard remains for a long time, although with less and less hazard as the time gets longer. The final question is whether they can be safely removed from contact with society for the long periods necessary for their decay to levels typical of normal background. The planned method is final deep burial in stable geologic formations perhaps 2000 feet down. Initial storage for up to 10 years may be in surface storehouses with adequate cooling, since the radioactive decay energy generates considerable heat. The rate of heat generation drops rapidly over the first 10 years as the short half-life elements decay away. Final storage is thus simplified by this moderate delay.

What assurance do we have that these wastes will not plague man in the future? First we have waste burial at several thousand feet deep. The problems encountered with toxic chemical wastes such as those from the infamous Love Canal near Niagara Falls, New York, have all been from shallow burial where leakage could readily contaminate local groundwater. Second is the form of the waste. The glass in which the radioactive material will be incorporated will be leached or dissolved only very slowly even if directly attacked by water. In flowing water at 40°C (104°F) only 1 mm (0.04 inches) of surface glass would be dissolved away in 100 years, for example. In any event the glass could initially be encased in stainless steel, possibly clad with lead or titanium acting as a barrier for thousands of years. If fuel elements are stored without reprocessing, their construction of zirconium and stainless steel plus encapsulation in a protecting canister assures extremely long lifetimes without serious degradation. A third important factor is that the highly energetic γ-emitting isotopes have relatively short half-lives, mostly less than 30 years. Thus after perhaps 600 years they will have essentially disappeared, and only the long half-life transuranics like plutonium or americium or protactinium will remain. These, however, are all α emitters that are only hazardous when actually ingested into the body. Because of their long half-life they are also weakly radioactive, and a person would have to ingest significant quantities to cause adverse health effects. One writer has estimated that even if a person swallowed or breathed a half-pound of radioactive waste that had been decaying for 600 years, he would still have a 50/50 chance of surviving the effects of the radiation.

A great many of man's technological advances have dangerous side effects. Pesticides can permit bountiful harvests but can gradually find their way to water supplies and pose long-term health hazards there. The concentration of lead in the atmosphere is increased by the exhaust fumes from automobiles, and the levels of sulfur dioxide and nitrogen oxides, both dangerous pollutants, are increased by the burning of coal in power plants. Many industrial processes produce wastes containing toxic chemicals such as arsenic, mercury, cadmium, or selenium. These do not decay with time but remain eternally as poisonous as when first liberated. The solution is not to ban the processes that produce these pollutants. It is to minimize their production and control their disposal. The EPA estimated that in 1980, 57 million tons of hazardous wastes were produced by the nation's manufacturing industries, and that 90% of it was disposed of in an unsound manner. Sound methods do exist for the disposal of these wastes just as they do for nuclear wastes. Neither one need be a plague for future generations if we are willing to put the known proper disposal technologies in place and pay for them.

The preceding discussion has been concerned with high-level wastes (HLW) and TRU. They are the direct products of the irradiation of nuclear fuel and require special care and handling — HLW because of the very high amounts of radioactivity involved, and TRU because of their very long half-life coupled with their α-emitting characteristics and the problems posed by possible entry into the body by eating or breathing. All other radioactive wastes are lumped in the category of low-level wastes for which less stringent disposal methods are required. About half the commercial low-level wastes are direct by-products of the nuclear power industry and the other half come from medical or industrial sources. Many of these, of course, may have originated as isotopes produced in nuclear reactors and specially separated for hospitals or industrial users. These wastes may include the ion exchange resin solids that have been used to purify radioactively contaminated water, equipment that has been used in intense neutron fields and thereby made radioactive itself, waste products from medical diagnostic tests, cleaning rags used to mop up radioactive spills, contaminated clothing, or even contaminated soil. They may be solid or liquid, bulky or consolidated. Most of it is contained in drums for shipping and for disposal by burial in shallow land trenches. Because of its polyglot origin it contains many different radioactive species with varying radiation characteristics and half-lives. Once encased in a drum, the radiation external to the drum can be measured and the method of handling can be appropriate to the radiation intensity. In most cases the use of conventional handling equipment such as cranes and lift trucks is possible, but on occasion special shielding for the operators may be desirable.

Disposal of low-level wastes has usually been in wide trenches where the drums may be covered with several feet of soil, enough to minimize the radiation at the ground surface. This is the same type of disposal as is used with toxic chemical wastes, and many of the same problems exist. The primary one is the extent to which leakage from faulty or corroded drums can seep out into the soil and eventually reach groundwater. Two characteristics of radioactive wastes make them easier to handle than toxic chemical wastes. First, their very radioactivity makes them easy to trace. The highly sensitive instrumentation today can detect even the tiniest amounts of radiation in water supplies, for example. The problem may be to distinguish between the radioactivity added by the leaking waste from that already naturally present in the groundwater. Revelation of the presence of toxic chemicals, however, may require relatively complex chemical analysis. Second, the radioactive wastes become continuously less and less potent and eventually subside to essentially background levels. With some radiation this may take a long time, but for most low-level wastes a few hundred years will suffice. Toxic chemicals may retain their toxicity forever.

Placed in proper perspective, the problem of disposal of low-level radioactive wastes blends into the overall problem of disposal of all of the hazardous wastes our society generates. We have not handled them well in the past because we have not fully recognized the long-term implications of inadequate disposal. As a society we have been too willing to throw our empty beer cans from the car window as we drive along or stash our rusted vehicles in the nearest patch of woods or set aside a few acres for a landfill into which anything may be thrown and lightly covered by the town bulldozer. The prime characteristics of our waste disposal have been that it must be easy and it must be cheap. Because of its emergence from a high technology industry, there has been more care taken with radioactive wastes than with most others. In some instances it has not been enough, and clean up problems remain as at the disposal site at West Valley, New York.

These problems are not because of inadequate technology, as critics charge. They are because the available technology was not properly used or because attempts to save money led to use of lower grade technology than was needed. There is adequate space available for disposal of our wastes. There is adequate technology available. We must be willing to see that it is used. We must be willing to pay the price required.

In December 1982 Congress passed the Nuclear Waste Policy Act of 1982 establishing a comprehensive long-term program for the disposal of high-level nuclear wastes. As called for by the law the Department of Energy in December 1984 recommended three sites and two alternates for consideration as the first of two permanent HLW depositories. The sites are Yucca Mountain, Nevada; Deaf Smith County, Texas; and Hanford, Washington. The alternates are in Mississippi and Utah. Further environmental assessments and detailed site characterizations will be carried out prior to presidential approval of a final single site scheduled for March 1991. Actual operation will not be before 1998. Recommendations for a second site will follow on a path aimed at operation about 9 years after initial operation of the first site. There are provisions for extensive public hearings to ensure environmentally acceptable sites and construction.

The federal government has thus committed itself to a safe and effective storage method for HLW and TRU by the late 1990s. The bill also provides for interim storage of spent fuel at one or more federal facilities if adequate private facilities are not available. The Department of Energy has selected Oak Ridge, Tennessee, as the site for such a monitored retrievable storage facility (MRS) and is preparing site-specific designs capable of handling both spent fuel and HLW. This MRS facility can be a place for temporary storage until the final repository is ready.

The costs of these facilities (except for interim storage that will be billed directly) is being paid for by a 1 mill/kWh tax on all electric power generated from nuclear plants after April 1983. The nuclear industry is thus paying for disposal of its own wastes, as should properly be the case.

In Chapter 10 we discussed the Low-Level Radioactive Waste Policy Act of 1980 and the amendments passed in 1985. These cover low-level radioactive wastes from all sources, including nuclear power, and will lead to adequate disposal areas for all states by 1996 at the latest. In the meantime the three functioning sites (Richland, Washington; Barnwell, South Carolina; and Beatty, Nevada) will continue to handle the nation's low-level nuclear waste.

Any new site chosen for either HLW disposal or state or regional low-level waste disposal will have problems with public acceptance. No one likes to have a dump in his backyard. Selection of sites will have to be made intelligently and carefully to minimize these problems. With proper management the sites can be made both safe and sightly. Local residents can and should insist on this. Our complex society cannot exist, however, without dumps, power lines, roads, railroads, and airports. They cannot always be on or next to someone else's property. Sometimes they must be our own neighbors. When this occurs we can insist they be engineered properly, but attempts to prevent them by force can only undermine the foundations of our democratic society.

WEAPONS PROLIFERATION

The third major objection to nuclear power stems from fear that the existence of nuclear reactors for power will lead to widespread availability of nuclear weapons. The rationale is that the spent fuel from these reactors can be reprocessed, either secretly or openly, and the resulting plutonium can be used to make weapons. This possibility was recognized in the early attempts to control nuclear weapons and led to the international Treaty on the Non-Proliferation of Nuclear Weapons sponsored by the United Nations and in force since 1970. The signatories of this treaty (about 110 countries) agreed to renounce the possession and use of nuclear weapons in return for access to information, equipment, and materials for peaceful uses of nuclear energy. They also agreed to inspection of their nuclear energy activities by the International Atomic Energy

Agency (IAEA), which was established by the United Nations in 1957. This inspection would in theory be able to detect any clandestine diversion of nuclear fuel to weapons.

Under the Non-Proliferation Treaty (NPT) those states already possessing nuclear weapons were allowed to keep them and add to their numbers at their own discretion. That included the United States, the USSR, Great Britain, France, and China. The treaty was an attempt to limit the spread of nuclear weapons to other countries and was in no way a control on the existing nuclear weapons powers. A number of major countries felt they had to keep the option of nuclear weapons open, and they refused to sign the treaty. This list includes Israel, Pakistan, India, Argentina, Brazil, Spain, and South Africa. India detonated a nuclear explosive "for peaceful purposes" in 1974 and thus clearly has weapons capability. There is some evidence that South Africa may also have tested a nuclear explosive, but this has not been confirmed positively. Other nations can drop their adherence to NPT on three months notice to the United Nations.

We thus have a situation where nuclear weapons exist in quantity in 5 countries including the United States and the USSR, which are the most likely to use them, and where several other countries may already have them or be close to having them. Most other nations have publicly renounced the intent of having or using such weapons. Does the expansion of nuclear power seriously change the probability of future weapons expansion? That is the question we have to answer.

The nations that already have nuclear weapons obtain the weapons material by one of two paths — either the separation of uranium isotopes to provide nearly pure ^{235}U or the generation of plutonium in reactors designed or built for that purpose. In a few cases such as the so-called N reactor at the Hanford, Washington, site in the United States, by-product power is produced, but the primary purpose is weapons-grade plutonium. The term weapons grade means that it is relatively free of the isotopes ^{238}Pu, ^{240}Pu, and ^{242}Pu, which are spontaneous neutron emitters and complicate the instantaneous triggering of the weapon. The objective of a commercial nuclear power reactor is to irradiate the uranium fuel as long as possible. The irradiated fuel elements removed therefore contain a relatively large amount (25 to 30%) of the isotopes that seriously lower weapons performance. The fuel in reactors designed for production of weapons-grade plutonium is removed much more frequently so that the separated plutonium can be 95% or better ^{239}Pu.

A nation wishing to enter the "nuclear weapons club" has three possible courses. It can openly declare its intention to do so and design and build isotope separation plants or plutonium-producing nuclear reactors. It can do the same thing secretly. It can divert spent fuel from its commercial reactors, reprocess it, and use this low-grade plutonium for its weapons. The first course, open declaration, was the route elected by France. Additional nations selecting that course today would risk losing the aid that the "Suppliers Club" (the group of nations posessing the technology, equipment, and materials needed for commercial nuclear power) could provide them. This would not be a serious loss to a nation already technologically sophisticated and with access to supplies of uranium, but it might be a significant loss to smaller and less strategically located powers. The third course, diversion of plutonium from commercial power plants, has several serious drawbacks. The plutonium obtained would be of poor quality for weapons. The diversion would have to be hidden from the inspectors of the IAEA. The amounts of plutonium obtainable without major risk of detection would be small.

It is the second course, clandestine establishment of reactors or isotope separation plants, which is the most likely source to be used. It would yield high-grade weapons material. The amount of material produced could be essentially unlimited. All of the external assistance for peaceful application of nuclear power could be preserved.

It seems highly probable, therefore, that the spread of nuclear weapons to other nations will be dictated by their national goals and objectives and will be unrelated to whether or not the nation has nuclear plants for the generation of power. The presence of nuclear power plants will not significantly influence the decision of such a nation to have or not to have nuclear weapons; even if the decision is in favor of weapons, the nuclear power plants will probably not be used as the source of weapons material. The possibility of spread of nuclear weapons to additional nations is a threat to all of us. It is a threat that will exist with or without the presence of nuclear power plants and should therefore not be used to block the growth of nuclear power.

In summary, the existence of nuclear weapons will depend on the political desires of the various nations and will be little, if at all, influenced by the presence or absence of nuclear power. Safe disposal of the wastes from nuclear power is technically feasible and now has a solid base in United States law. It is undeniable that there is some risk from an accident at a nuclear power plant. The risk is associated only with the most severe accidents, however, and these have very, very low probabilities of occurrence. Even in their most extreme form these risks are comparable in magnitude to others we accept as reasonable, and new technical analysis seems to indicate that the maximum possible harm may be less than we had previously thought. Thus the risks from nuclear power truly are very low. What are the rewards that must offset them?

REWARDS

To consider the rewards of nuclear power we have to consider the whole question of energy in our society, why we need it, and where it comes from. Prior to the nineteenth century man's own labor was augmented largely by the labor of animals drawing on fuel from agricultural feed, by wind power from the windmills like those in Holland and on U.S. farms, and by waterpower from rivers used to drive mill wheels and other mechanical devices. Concentrations of population in towns and cities had to be self-sufficient or draw on nearby farm areas for food. Commodities such as oil for lamps or wax for candles were of animal or vegetable origin, and the metals of commerce were smelted and forged with coal or wood as the primary energy source. The energy infrastructure was relatively simple and depended largely on local supply.

The inventions of the nineteenth century—the steam engine and electric light and power and the discovery of petroleum and gas as major energy sources—provided the basis for a profound change in the way man used and obtained energy. This extraordinary pace of change was accelerated further in the early twentieth century by the development of the automobile and airplane as enormous users of energy providing rapid and convenient transportation. Equally important, our cities have become almost totally dependent on electric power for their continued functioning. One has only to recall the chaos that ensued in the 1977 electrical failure in New York City and which resulted in a long total blackout. Lights, cooking, elevators, traffic control, subways, industrial machinery, and many other facets of the city's functioning existence depended on the continuity of electric power. Its loss and the looting and temporary anarchy that followed dramatically indicated how much our cities depend on a secure and economic source of electricity.

This secure and economic source of electricity is the reward to which nuclear power contributes. In 1984, 13.5% of the nation's electricity came from nuclear plants, and by 1990 this figure is expected to grow to nearly 20%, assuming completion of most of the plants now under construction and scheduled for completion by then. The generation of the enormous amounts of electric power required to sustain our cities, towns, and major industries requires the consumption of primary fuel. Those available in adequate quantities are oil, gas, coal, and uranium. Oil is also the preferred primary fuel for transportation. It is already approaching the stage of serious depletion, and its use for electric power generation is clearly unwise. Gas is an ideal fuel, but it is also in limited supply, although very deep well drilling may improve the supply situation to some extent. In 1980 approximately 15% of our electric power was generated by burning gas, but by 1983 this had dropped to 11.9%, and by the year 2000 is expected to be as low as 5%. In the absence of nuclear power this would have to increase substantially, but it is doubtful that adequate supplies would exist to provide more than 10 to 15% of our electricity needs.

Coal is the one energy source besides uranium available in adequate supply for our energy needs for the next few hundred years. Currently nearly 50% of our electric power generation comes from coal. It is far from an ideal fuel, however. A 1000 MW coal-fired electric power plant will consume about 2.5 *million* tons of coal per year. This compares to a mere 33 tons of slightly enriched uranium for a nuclear plant. The relative magnitude of the fuel transportation problem is thus easy to see. Coal has gaseous emission and waste disposal problems also. The burning of coal generates carbon dioxide (CO_2), sulfur dioxide (SO_2), and nitrogen oxides (NO, NO_2, etc.). CO_2 is not a poisonous gas, but there is some scientific concern that too much CO_2 added to the atmosphere can cause changes in the way in which our atmosphere permits entry and exit of incoming and reflected solar radiation. The net result might be a warming of our atmosphere sufficient to cause melting of polar ice caps and a rise in the level of oceans sufficient to flood some of our coastal cities. You don't need to start evacuating now, because the so-called "Greenhouse Effect" will occur very slowly and gradually, and we are not even positive it exists. Sulfur dioxide and nitrous oxides are mildly toxic, and in sufficient quantities can be deadly. It was largely very high SO_2 levels from burning soft coal that caused nearly 4000 deaths in a several-day period in London in 1952. An unusually stagnant air mass permitted the high SO_2 buildup to levels that were highly toxic to older people or those with chronic lung problems. SO_2 has also been implicated as a main source of the so-called acid rain that has caused damage to fish life in ponds and streams, particularly in the Northeast portions of the United States.

Large quantities of ash are left as the residue from burning of coal. A 1000 MW plant will produce from 600 to 700 tons of ash per day, so a typical plant must bring in 6500 tons of coal per day and take away 650 tons of ash. Much of this ash would be carried up the great chimneys of such a power plant and distributed over the surrounding landscape were it not for the electrostatic precipitators that remove the tiny ash particles from the hot gases carried up the stack.

SO_2 control is a more difficult problem. It can be attacked by use of selected low-sulfur coals (at premium prices), by chemical cleaning of the coal before burning, or by complex and costly chemical "scrubbing" of the exhaust gases escaping from the chimney or stack. Unfortunately the use of low-sulfur coals makes ash removal from the stack more difficult, because the presence of sulfur in the exhaust gas makes the electrostatic precipitators work more effectively. All of these control methods are technically feasible, although they have not been universally applied as yet and will considerably increase the cost of electricity production from coal as they become more commonly used.

Our arguments so far have shown the need for electric power and that there is insufficient oil and gas for them to provide the power essential in the future. We have also shown that there is ample coal and uranium to fuel our energy needs long into the future, although both fuels have side effects which, while controllable, pose measurable risks. We must still consider the possibility of supplying all the electric power we need from clean, renewable sources such as solar energy, geothermal, wind, and water power. There is a scenario, believed by many, that by combining careful conservation with extensive development of these renewable sources of energy we can avoid the use of either coal or uranium for electric power production in the future.

This is an enormously appealing concept, but it must be viewed with detached realism. Let us consider conservation first. Prior to 1973 the growth of electric power had been a steady 6 to 7% per year for over 50 years. This steady growth was based on ready availability of fuel sources and continuously decreasing costs. The world oil shortage in 1973 and the accompanying dramatic increase in the cost of oil and other energy sources brought a halt to our unchecked energy consumption and ushered in an era of complex changes in energy use patterns. The extraordinarily high interest rates of the 1970s and early 1980s contributed significantly to the major increases in the cost of large power plants, particularly for nuclear power where delays induced by stricter regulations, lower demand levels, and antinuclear harassment substantially lengthened construction time of nuclear plants. For the first time, therefore, electric utilities had an incentive to restrict the rate of load growth to escape the financially crippling impact of the huge costs of new power plant investment. After an initial period of shock, they eagerly embraced the concept of energy conservation to at least temporarily defer the need for new plants and to minimize the drain on oil and gas supplies now seen as in serious jeopardy.

Conservation has many aspects. To the homeowner or commercial building operator it means better insulation to preserve heat, living at lower but still healthful temperatures, turning off lights when not needed, designing homes to take advantage of solar heat, using electricity at times other than peak demand periods, etc. To the manufacturer it means investing in more efficient electric motors, improving the efficiency of processes using heat, or installing more efficient lighting systems. To the driver it means buying smaller and lighter cars and keeping to 55 miles per hour to conserve fuel. To the airline industry it means improving the fuel efficiency of engines and flying at optimized speeds and altitudes.

These are all ways of saving energy, but you will notice that only some of them involve saving electric energy. The greatest thrust has been towards the reduction in use of gasoline, fuel oil, and to a lesser extent natural gas. Between 1973 and 1982 in the United States the use of nonelectric energy *decreased* by a total of 15%. The use of electric power over the same period *increased* by 20%. In previous decades this increase would have been 60 to 65%, so conservation has substantially reduced the rate of growth, but it has not eliminated growth entirely. Nor should it. It is only common sense to use our electric power efficiently. It is equally common sense to use it productively. For many years there has been a direct link between the growth of our gross national product and the growth in use of electric power. Even since 1973 this linkage has persisted, granted with a slight change in the ratio, but still directly linked and demonstrating the dependence of GNP on electric power supply.

Electricity is such a safe, convenient way to transmit and utilize energy that its share of total energy consumption has increased consistently ever since its first use in the nineteenth century. It represented one-fifth of the energy used in the United States in 1960, one-fourth in 1970, and one-third in 1980. By the year 2000 it will probably be at least 40% of our total energy use. With continued growth of population and GNP and with major efforts to promote the highest degree of conservation consistent with reasonable prosperity, the use of electricity must inevitably continue to grow. We cannot predict exactly whether it will average 2% per year or 4% or even 6% per year, but it will grow.

Thus the demand for electric power cannot be answered by conservation alone, although conservation must play a part. What part can renewable resources play? Here we have to evaluate a number of different sources — particularly solar, geothermal, wind, and water. The total flow of radiant solar energy to the earth is 10^{18} kilowatt hours (kWh) per year. The world's energy needs today are perhaps 6×10^{14} kWh/ yr, so there would appear to be more than a thousand times as much energy coming from the sun as we need. The problem lies in our ability to concentrate it where and when we want it. A city like New York, for example, needs huge amounts of power compared to the needs of farmland in rural areas. This power need is greatest in the middle of winter when the amount of solar radiation is least even on sunny days and when the number of cloudy days compared to sunny ones is the greatest.

Completely unrealistic amounts of space would thus have to be utilized for solar power to become a major source of electric power generation for such a city.

Solar power can and should play a larger role than it currently does, however, so let's consider how and where it can be useful. First, solar power can be converted directly to electricity (DC) by the use of photovoltaic cells. These cells, made from highly purified silicon or other semiconducting materials, have relatively poor conversion efficiencies (10 to 20%) and are still expensive. They are extremely useful where small amounts of power are needed and where they can be coupled to battery or other storage systems to assure continuity of power at night or on cloudy days. They have not yet been shown to be practical for generation of the large blocks of continuous power required for urban areas. The largest photovoltaic electric power generation system built to date, for example, is a 6.5 megawatt (MW) plant built by Arco Solar Incorporated in California. The DC power produced is converted to AC and is sold to Pacific Gas and Electric for use in their system. The period of maximum power generation coincides with the hottest period of the day when the system's air-conditioning load is the greatest. There is thus a good match between the solar electric power production and the peak system demand. The system base load must still be provided by conventionally fueled sources, however.

It is significant that this plant and a similar 1 MW photovoltaic plant Arco built to supply power to Southern California Edison are owned and financed by Arco. By federal law the utilities are required to pay third-party power producers a price equivalent to the utilities highest avoided cost (that is, the most expensive power that can be replaced). This is usually the power from the old oil-fired units. This amounts to a subsidy for solar power to encourage its broader utilization. Such a third-party plant is only permitted by law to produce 80 MW power— not much when compared to a 1000 MW coal or nuclear plant.

It is clear that solar photovoltaic electric systems have a suitable and perhaps growing role to play as a source of power. A number of basic factors such as the land area required, system cost, and the timing of power availability will prevent them from ever providing more than a minor share of our electric power.

Similar factors will limit the utility of solar thermal systems for electric power generation. In these systems fields of reflecting mirrors focus sunlight onto a thermal receiver attached to a boiler system generating steam to drive a turbine generator and produce AC power. The largest such system is an experimental 10 MW plant called Solar One at Barstow, California. Here computer-controlled mirrors focus the sun's energy onto a receiver atop a 310-foot tower where a boiler and turbine generate power for the Southern California Edison grid. A 100 MW solar thermal plant is planned for operation in 1988, but costs are estimated to be marginal compared to gas or coal-fired plants. Approximately 17 acres of reflectors are required per megawatt of electric power capacity, so a 100 MW plant would require 1700 acres of mirrors.

California, Arizona, and New Mexico will see increasing use of photovoltaic and thermal solar energy, but even there the rate of growth will be relatively slow because of marginal economics. The northern and eastern parts of our country will embrace solar electricity even more slowly, if at all, because of far less favorable conditions in these regions. Most of the northern utilities are winter peaking, that is, their maximum customer demand comes from the need for heat at the coldest period of the winter. This is when the available solar radiance is least, just the opposite of the peak demand in southern California for cooling during the hottest part of the summer. In Boston, for example, the average daily radiation in December is 443 British thermal units per square foot (443 Btu/ft^2). This compares with an average daily radiation level of about 2400 Btu/ft^2 in July in Los Angeles. So the energy available in Boston in December is less than one-fifth as much as in Los Angeles in August. Add to this the fact that in Boston or New York or Chicago it is not unusual to go many days without substantial stretches of sunshine, and you can begin to see some of the problems with solar electricity as the primary source of power for these regions.

Solar radiation is also the basic driving energy behind the winds that blow constantly across our country. This energy can be harnessed to provide electricity via windmills, or wind machines as some people prefer to call them. The old multibladed farm windmill has been converted by modern technology into a variety of far different and more efficient machines to drive generators to produce electric power. Single wind machines have been built generating as much as 4 MW power, and larger machines up to 7 MW are under development. A typical 3 MW machine has a single enormous propellor blade 260 feet in diameter. It must be designed to produce useful power in as light winds as possible yet withstand the most violent winds occurring over a 30 to 40 year life span. These are not trivial design problems, and the developmental models have had their share of problems as well as successes. It is sobering to recognize that it would take over 300 of these largest machines placed in unobstructed locations (probably of high visibility) to generate power equivalent to that from a single large nuclear or coal-fired station. In addition these machines would only produce power when wind conditions were favorable, neither too light nor too strong. Southern California Edison as of 1982 had established a goal of 560 MW wind capacity on their system by 1990. This would be equivalent, in their estimation, to only 140 MW "firm capacity" because of its intermittent availability. Thus wind power, also, is no panacea. It should and will provide modestly increasing amounts of power, but at high capital cost and uncertain reliability.

The natural heat of the earth is another source of thermal energy convertible to electricity. Naturally occurring steam at the Geysers near San Francisco has for many years been fed into turbine-generators by Pacific Gas and Electric and converted into electric power for their grid. As far back as 1979 they had a capacity of 545 MW, and the possibility exists for as much as 5000 MW from this single field. The Geysers field is apparently unique in the United States; no other similar sources are known. It is possible to extract power from natural hot brines existing underground in Texas and Louisiana, but these are highly corrosive materials and their practical use has not yet been demonstrated. Brines in Mexico have been utilized successfully, however, so the concept is technically feasible. Natural hot water sources or geological areas where rock temperatures underground are as high as 200 to 300°C can also be exploited in the future. Techniques such as these are limited in use to the areas where these natural conditions occur and are not universally possible across the nation. Their economics are not clearly favorable, and the question of disposal of the brines after their use to generate steam poses a problem. The solution generally proposed is reinjection into the earth where they came from, but because of the pressures involved this is not a simple procedure. The National Research Council of the National Academy of Sciences concluded as follows in its 1979 report *Energy in Transition 1985–2010:*

> Considered in all of its potential, the geothermal resource represents extremely large amounts of energy. However, for a variety of technical, economic, geographical and institutional reasons geothermal energy will probably not be a major contributor to the national energy system until well into the twenty-first century if ever.

Anyone who has seen the power and beauty of Niagara Falls or has watched the water cascading down the face of Grand Coulee Dam is aware of the potential of water power for the generation of large amounts of electricity. Grand Coulee alone has a capacity of 6430 MW, Churchill Falls in Canada generates 5225 MW, and the new Brazilian site at Itaipu will eventually generate 12,600 MW. There are not many sites like these; in the United States in particular all the obvious ones have been already exploited. As early as 1920 there were 4000 MW hydropower in the United States, and by 1981 this had grown to 62,000. But other fuel sources for electricity had grown much faster in the interim, and while in 1920 hydropower generated 31% of all the electricity produced, by 1983 the percentage had dropped to only 14.4%. In spite of continued development of new sites, this percentage is projected to drop to perhaps 9% by 1990.

At favorable sites hydro is abundant and cheap, but the major U.S. sites have already been exploited and only less desirable ones remain. These are in streams with smaller flows and lesser drop heights, so the capital costs per kilowatt of generating capacity tend to be higher. The environmental consequences of damming up such streams are frequently considerable in terms of land flooded or wildlife displaced. Nevertheless hydro projects are attractive where modest blocks of power can be utilized, and small hydro installations are on the increase on U.S. streams. Their total impact cannot be large; while they will reduce the need for coal-fired and nuclear plants, they cannot replace them.

All the renewable sources such as solar, wind, geothermal, and hydropower must be vigorously pursued under the conditions where each is favorable. Even together they cannot provide the amount of electric power our complex society must have to function effectively now and in the future. Nor can conservation, however stringently applied, relieve us from the necessity of supplying our society with growing amounts of electricity far into the future.

The dilemma we face is thus a choice between two competing technologies, coal and nuclear, each with serious natural drawbacks in potential ecological damage but each subject to technical safeguards to minimize their effects. The reward we obtain from their use is the ability at rational cost to light and heat our homes and workplaces, to power our factories, and to enable the use of a myriad of labor-saving and entertainment devices that enrich our lives. The reward of cheap electric power is enormous. It is deeply ingrained in our daily life and is absolutely essential to the survival and prosperity of our cities. The pivotal question in regard to nuclear power is whether the tiny, tiny risk of a massive accident with many deaths and up to billions of dollars in damage is a risk worth taking to ensure the continuation of this enormous reward we too often take for granted. Were it not for the existence of coal, as far as the United States is concerned there would be no question but that the reward was more than ample for the very small probability of having to pay a large but not uniquely large price. By using coal we substitute a much greater certainty of continuous moderate levels of risk in mining the massive amounts of coal needed, in transporting it, in disposing of the huge tonnage of waste ash, and in the cumulative effects of emission of large amounts of carbon dioxide, sulfur dioxide, and oxides of nitrogen.

A strong case can be made in support of both coal and nuclear. We currently mine about 600 million tons of coal a year. Total future dependence on coal in the United States would require production to reach perhaps 2 billion tons by early in the next century, a disturbingly disruptive amount. The safety record of nuclear power has been astoundingly good. The only question is whether we can keep it that way. We must continue to refine and perfect our estimates of what the real risks are. We must reconsider our siting criteria and perhaps the maximum size of individual units to be built. We must continue to enhance the safety of existing and future nuclear plants by design and operating improvements. We cannot afford to abandon the nuclear path because the rewards are societal stability itself; there is no clearly better alternate and the risks, although present, are acceptable.

EPILOGUE

FROM THE SPACE SHUTTLE a few hundred miles up, the earth can be seen as a beautiful blue sphere floating serenely in an immense atmosphere and bathed in sunlight. The astronauts know that in addition to sunlight there are cosmic particles of immense energy traversing that otherwise empty space and hurtling into the earth's outer atmosphere, there to form cascades of secondary particles eventually reaching the earth and all its inhabitants in many forms. Other wavelengths of electromagnetic radiation from the sun and beyond irradiate all the objects in space including our earth. In a tiny way the shuttle itself adds to the spectrum of radiation through microwave signals to its ground-support crews.

Our world has existed in this sea of radiation for billions of years, and the evolution of life and of man himself has occurred in this radiant atmosphere and possibly in part because of it. Now man is adding variety and changing the local intensities of many kinds of radiation to provide communication, power, diagnostic information, and healing. Along with these benefits our growing understanding of all aspects of radiation has made us aware of the hazards coexisting with the enormous benefits radiation provides. Perhaps because we see them dimly, they have frightened us too much. Perhaps enamored with benefits gained, we have sometimes overlooked the harm.

The key to wisdom is understanding. Radiation is basically neither good nor evil. It is a phenomenon as ancient as time. It exists for us to use and to control. If we understand it, we can control it; and if we control and use it wisely, we can truly speak of our radiant world.

CHERNOBYL

ON PAGE 186 OF THIS BOOK you read the sentence, "We can only point to the 29-year history of commercial nuclear power to date with no record of any direct consequences at all to human health." That was true when written, but on 26 April 1986, after this manuscript had reached page proof, the situation changed dramatically. At 1:23 AM on Saturday, 26 April, the worst accident in the history of commercial nuclear power was initiated. It has resulted in deaths and injury, and we must try here, albeit before all the facts are available, to place the accident at the Chernobyl power station north of Kiev in the Soviet Union in rational perspective with what has been said in this book.

To do this we will discuss first the status of nuclear power in the Soviet Union and the main design features of the Chernobyl reactors. Then we will cover the nature and magnitude of the accident as far as the details are known at this writing (26 May, just one month after the accident), and finally we will consider the long-range consequences of the disaster and its significance for the future of nuclear power.

While commercial nuclear power in the United States can be said to have begun with the initial operation of the Shippingport PWR in 1957, the Soviet Union claims 1954 as the origin of commercial power through the initial operation of their 5 MW graphite-moderated, water-cooled reactor at Obninsk near Moscow. This was the prototype for a series of similar reactors of steadily increasing size culminating in the enormous 1500 MW reactors at Ignalino (one started up in late 1984, the other planned for 1986) and Kostroma (scheduled for the end of 1986). Parallel to these reactors, but beginning about 1964, the Soviet Union developed also a series of pressurized water reactors (PWRs) generally similar to the design of U.S. and French PWRs. While recent commitments have been more towards the pressurized water type, the graphite moderated design (designated RBMKs) are still being built. In 1982 almost 64 percent of the operating Soviet nuclear plants were RBMKs, and they had a total output of 10,000 MW. Prior to the Chernobyl accident, plans existed for attaining a total electric power capacity by 1990 of over 40,000 MW via PWR-type reactors and at least an additional 21,000 MW through RBMKs. For comparison the planned United States total nuclear capacity in the same period is about 119,000 MW.

The reactor that was destroyed at Chernobyl was an RBMK-1000, a 1000-megawatt version of the graphite-moderated, light-water-cooled design. It was the most recent of a series of four 1000-MW reactors of that type at the Chernobyl site. These reactors are quite different from the boiling or pressurized water reactors that are the mainstay of nuclear power in the United States. They consist of a very large graphite core made up of 250-millimeter (about 10-inch) square columns of graphite. There are 2488 such vertical columns making up a cylindrical core block 12.2 meters (40 feet) in diameter and 7 meters (23 feet) high. Through the graphite pass 1661 tubes made up of zirconium alloy in the central part and alloy steel at each end. These pressure tubes are 88 mm (3.5 inches) in diameter with 4 mm wall thickness and contain the zirconium-clad uranium oxide fuel elements arrayed in assemblies over which water passes to extract the fission heat and generate steam. Interspersed with the fuel elements and parallel to them in the graphite are 211 movable control rods containing boron carbide that control the reactor power. The steam generated in each pressure tube passes to a header in which all the steam is combined and is then routed via a steam separator to the turbines and generators producing the electric power that is the plant's product.

The fuel is inserted in and removed from the reactor from above, using a fuel-loading machine and a large crane that are housed in a tall building extension directly above the reactor core. There is heavy concrete shielding surroundng the space immediately above the reactor, but in the uppermost part of this central upward building extension the design is lighter. This is a critical point, since it bears on the issue of whether the RBMK reactors have a containment building in the same sense as the PWR and BWR reactors in the United States. As this is written the design containment pressure of this portion of the RBMK reactor has not been indicated. It appears to be significantly lower than for U.S. reactors, however.

With these basic design facts in mind, we can now consider the nature of the accident itself. The complete sequence is not known at this writing, but the basic facts appear to be these. At the beginning of the incident the reactor had been reduced to about 20 percent of full power, presumably on the way to shutdown. At that point either there was a mistaken withdrawal of one or more control rods permitting a sudden large burst of power, or there was a blockage of coolant flow permitting rapid overheating of significant amounts of fuel. In either event the fuel and zirconium clad reached very high temperatures at which the zirconium reacted with steam to generate hydrogen. It is possible also that ruptures of some of the pressure tubes permitted steam to contact the graphite moderator, also at high temperature, and generate carbon monoxide (CO), hydrogen, and methane (CH_4), all combustible gases. The reactor core is believed to be surrounded by a gas-tight shell within which an inert atmosphere of nitrogen and helium is contained. Thus under normal circumstances any combustible gases generated could not explode because no oxygen would be available for combustion. It seems clear, however, that somehow the gas-tight shell was ruptured so as to permit escape of the hydrogen and possibly CO and CH_4 into the air space above the reactor and its top shield. These gases would naturally rise to the highest portion of the building where the crane and refuelling apparatus were housed.

The above scenario may not be exactly correct, but it seems fairly certain that in some way large amounts of combustible gases including hydrogen did collect in the air space of the main building and were ignited to cause a major explosion. The upper portion of the building was destroyed, probably dropping the crane and fuel-withdrawal equipment onto the top of the reactor shield so no fuel could be removed. Probably at this point also a major fire broke out involving, at least in part, the graphite moderator of the core. Dense graphite, such as in this core, requires significant amounts of air to burn, but air access may have been provided by the chemical explosion or perhaps by any disruption caused by the nuclear power excursion. Its burning mode would in any event be the slow advance of a glowing front as air is supplied, reminiscent of the charcoal fires on one's outside grill. This could cause further involvement of the fuel in the remaining zirconium pressure tubes, however, and lead to continued release of radioactive fission products.

The explosion and its aftermath fires must have left some terrible problems. We do not know at this point how much cooling capacity had been lost in the core, whether the control room was still operating, and what the radiation levels were throughout the reactor building. They were clearly very high as was shown by the number of cases of radiation sickness resulting. The reactor operators had to cope with escaping steam as well as radiation, try to bring the situation under control, and at the same time assure the safety of the other three reactors on the site. Not an enviable task. The details will emerge in time, and it does not pay to speculate further on them here. The eventual solution was to bring in helicopters to dump tons of sand, boron, and lead pellets onto the exposed top of the reactor to contain as much of the radiation as possible; smother the graphite fire; and ensure, via the boron, the stopping of any further fissioning within the remaining reactor fuel. By 1 May, five days after the initial explosion, the graphite fire appears to have been smothered and temperatures throughout the reactor had dropped to levels giving assurance that the situation was under control.

Two workers were killed in the initial explosion, 1 by exposure to hot steam and the other struck by flying debris. Many other workers including firemen and medical help were exposed to very high radiation levels. In all, 299 people were hospitalized, at least 35 with critically high radiation doses. Thirteen of these were given bone marrow transplants and 6 were given fetal liver transplants to try to restore the functioning of essential blood cell generation. By 16 May, 11 of the critical 35 had died, with expectations that others would die also. At this point it is believed that all of those hospitalized and all deaths were from the group of people working in the plant at the time of the accident or the emergency fire and medical help brought into the plant.

At the time this is written (26 May) less is known about possible harm to the people residing in the town of Pripyat where the reactor was located. This town was built to house the workers at the nuclear plant. The average age of the residents was 26. Mayor Voloshko was quoted in *Soviet Life* in February 1986 as saying that Pripyat was currently experiencing a "baby boom." The fallout radiation from the plant thus fell where a significant number of small children and pregnant mothers lived.

The extent of this fallout is still very unclear as this is written. The explosion on 26 April, which totally breached whatever containment existed, spewed out a pulse of radioactive gas and debris high into the air. While some of it dropped into the town surrounding the plant, since it was not raining at the time, much of it rose high in the air and drifted with the wind towards the northwest. Most of the population was presumably asleep in their homes at the time of the explosion. Did they stay inside, or, awakened by the explosion, did they make the mistake of running out to see what had happened? How much of the fission product burden of the plant was released in the initial explosion and how much later as a result of the fire in the graphite? Some U.S. estimates put the total as high as 50 percent, but it may have been much less.

Either the extent of the early fallout in Pripyat was not sufficient to trigger immediate evacuation or the confusion that existed prevented early organization of such an evacuation. Continued emissions from the stricken plant and greater recognition of the hazard in the surrounding area finally resulted on 27 April in a major evacuation of an 18-mile area surrounding the plant. Over 1000 buses were brought from Kiev, and working door to door effected the removal of 40,000 people in a reported 3 hours. In the ensuing week additional people were evacuated and the final number reportedly reached 92,000.

The radioactive dispersion in the atmosphere from Chernobyl took an erratic path in the days after the accident. First it drifted to the northwest towards the Baltic Sea, but in a relatively rainless period. It touched Sweden and Finland late in the second day. Then wind changes to the south and west spread the cloud out considerably and moved some of the continuing emissions from the plant over Kiev. There were no heavy rains during the early period of cloud drift, but some rain and fog affected the fallout pattern about 72 hours after the first radioactive release. There was thus a relatively broad area in the Soviet Union and some parts of Sweden, Finland, and Poland where fallout radiation occurred. None of the amounts appear to have been sufficient to result in health effects but caused concern about possible crop impact and led to embargos by other nations on produce from the affected areas. Thus serious fallout seems to have been confined to the Soviet Union itself.

While speculation at this early stage is hazardous we will risk the following tentative conclusions:

• The external radiation levels received by the people in Pripyat in the period between the explosion and their evacuation were under 50 rem and in most cases probably much under. This is based on the facts that there is no evidence that many of them were hospitalized for symptoms of radiation sickness, that no immediate evacuation was ordered, and that much of the cloud appears to have risen quite high.

• Inhalation of ^{131}I and its concentration in the thyroid may cause thyroid nodules, some cancers, and some deaths in the long term in the residents of Pripyat who were evacuated. Most at risk will be young children. There are not enough facts now to permit quantitative estimates of the extent of the problem.

• There will be no deformed children born as the result of the accident. Hiroshima evidence showed that levels of 50 rem whole-body radiation were required for this to occur, and these levels should not have been accumulated by pregnant women in the area.

• There will be no genetic aftereffects from the accident.

• There will be massive cleanup costs both around the reactor and in the areas of most intense fallout, particularly the 18-mile evacuated zone, which may not be resettled for some time.

• Care will have to be taken to monitor all dairy products from the region for several months to make sure it does not contain excessive amounts of ^{131}I. Sale of milk from critical areas must be banned. Vegetable produce and regional meat will have to be carefully watched to eliminate any containing undesirable amounts of ^{137}Cs, ^{90}Sr, or other isotopes.

• Cancer rates in the fallout area will probably show some increase over the years to come. The amount is hard to predict, but increases of 5 to 10 percent in cancer rate would not be surprising.

Even without the final accounting of damage it is clear that Chernobyl was the worst nuclear accident in history. Its acknowledged death toll is now 13 and will presumably go higher. The reactor itself is a total loss and will be entombed permanently in concrete. 92,000 people have been evacuated. A measurable, but not enormous, increase in thyroid abnormalities, cancers, and cancer deaths will presumably occur in the future. The toll will be much less than from the chemical accident at Bhopal in 1984, however, and the cancer increases far less than those caused by cigarette smoking. The casualities and damage are within the scope predicted by the Rasmussen report, but the occurrence of the accident at this time raises questions concerning the predicted probability rate for such accidents. We will have to await better understanding of the cause of the accident before making judgments on this point.

The Soviet Union will have to assess the significance of the accident for its own nuclear power program and the future of the RBMK reactor design. We in the United States must also consider the implications of the accident for our own nuclear program. We can take solace in that our reactor designs are fundamentally different than Chernobyl and thus the same sequence of events cannot happen here. We can point out that at Three Mile Island the containment concept worked and no one was injured, even in the plant itself. We cannot be complacent, however, because certain aspects of the Chernobyl disaster raise critical questions about all reactors.

First is the question of the generation of combustible gases and their role in possible chemical explosions that can nullify the advantages of containment. It was the *chemical* explosion of hydrogen or other combustible gases that turned the Chernobyl accident into a disaster. Can such a thing happen in our PWR and BWR designs? There is no graphite in our reactors, but we know that hydrogen generation from the high-temperature interaction of zirconium and water can occur in case of gross fuel overheating. It occurred at Three Mile Island, and the hydrogen generated became a source of concern there. It was concluded then that a combustible mixture and explosion could not occur. We must doubly validate that conclusion, triply test it, and fix permanently any design doubts such a study may raise.

Second is the question of reactor size, power density, and accommodation to error. There is much that says today's reactors are larger than desirable based on their total inventory of fission products. Can we design smaller reactors with lower power densities and 10- to 100-fold lower probability of major accident? Thus even though we will need more of these reactors to achieve our power goals, the possibilities of accident per 1000 MW power will be less and the consequences of the worst possible accident will be far less also. We must seek to make such reactors economical through standardization, modular construction, and unimpeded construction schedules.

Prior to the Chernobyl accident we believed our PWR and BWR reactors to be adequately safe. Nothing in this accident should change that conclusion unless reevaluation of the hydrogen generation possibility reaches different conclusions than before. We should therefore continue to operate these reactors as carefully and competently as possible. We should complete and operate those reactors currently under construction. We should also reassess our design basis for future reactors and mount a major national effort to develop a smaller, safer, standardized design with which to revitalize our needed nuclear power industry.

SELECTED CRITICAL
BIBLIOGRAPHY

THE FOLLOWING BIBLIOGRAPHY provides a listing of some of the major sources used in *Our Radiant World* as well as recommended additional reading. An asterisk* indicates references considered broadly useful and authoritative.

SOURCES AND EFFECTS OF NUCLEAR RADIATION

*Brill, A. Bertrand, ed. *Low-level Radiation Effects: A Fact Book.* New York: Society of Nuclear Medicine, 1982, 138pp.

Excellent loose-leaf notebook style volume containing brief text sections and many fine charts with selected data on use of radiation in medicine and biological effects of radiation.

*Committee on Federal Research on Biological and Health Effects of Ionizing Radiation. *Federal Research on the Biological and Health Effects of Ionizing Radiation.* Washington, D.C.: National Academy Press, 1981, 169 pp.

Sometimes known as FREIR Report. Report is an evaluation and critique of federally supported research effort on biological and health effects of radiation. It analyzes both research being done and the state of our knowledge resulting from it.

Goffman, John W. *Radiation and Human Health.* San Francisco: Sierra Club Books, 1981, 908pp.

This enormous and controversial book tries to make the case that low levels of radiation are substantially more harmful than estimated in the UNSCEAR or BEIR reports. Difficult reading but represents extreme view well.

Hubner, Karl F., and Shirley A. Fry, eds. *The Medical Basis for Radiation Accident Preparedness.* New York: Elsevier/North-Holland Press, 1980, 545pp.

A symposium organized by the Radiation Emergency Center/Training Site of the Medical and Health Sciences Division of Oak Ridge Associated Universities. International in scope, the papers cover the history, both actual and clinical, of many major accidental radiation exposures. Intended primarily for physicians, health physicists, and emergency personnel.

Kathren, Ronald L. *Radioactivity in the Environment: Sources, Distribution, and Surveillance.* New York: Harwood Academic, 1984, 397pp.

A graduate-level text discussing natural and man-made sources of radioactivity and emphasizing methods of measurements and surveillance.

*National Academy of Sciences, Committee on Biological Effects of Ionizing Radiation. *The Effects on Populations of Exposure to Low Levels of Ionizing Radiation.* Washington, D.C.: National Academy Press, 1980, 524pp.

Report of National Academy of Sciences Committee on Biological Effects of Ionizing Radiation, frequently referred to as BEIR III. Covers principles of effects of radiation and discusses in detail evidence for genetic, cancer, and other effects from low levels of radiation. Authoritative and readable.

_____. *Tritium in the Environment.* NCRP Report 62, 1979, 125pp.

Discusses the origins of tritium in the environment, its distribution in nature, and its biological behavior in animals and man.

National Council on Radiation Protection and Measurements. *Influence of Dose and Its Distribution in Time on Dose-Response Relationships for Low LET Radiations.* NCRP Report 64, 1980, 216pp.

Detailed analysis of effects of dose rate and total dose of X rays, gamma rays, and beta particles in inducing genetic or cancer effects at relatively low doses.

_____. *Critical Issues in Setting Radiation Dose Limits.* Proceedings of the Seventeenth Annual Meeting of the National Council on Radiation Protection and Measurements, Washington, D.C., 1981. NCRP Proceedings 3. 1982, 287pp.

Series of technical papers on risk assessment and its relation to establishment of realistic standards for permissible quantities of low-level radiation.

_____. *Environmental Radioactivity.* Proceedings of the Nineteenth Annual Meeting of the National Council on Radiation Protection and Measurements, Washington, D.C., 1983. NCRP Proceedings 5, 1983, 284pp.

Series of papers on origins of radioactivity in world today from natural sources and from man-made ones such as weapons tests and nuclear power cycle.

Organisation for Economic Co-Operation and Development, Nuclear Energy Agency. *The Environmental and Biological Behaviour of Plutonium and Some Other Transuranium Elements.* Paris, France, Sept. 1981, 116pp.

Discusses sources of plutonium and transuranics in environment, how they can be incorporated in the body, and what their effects are.

Scott, George P., and Heinz W. Wahner, eds. *Radiation and Cellular Response.* Report of the Second John Lawrence Interdisciplinary Symposium on the Physical and Biomedical Sciences held June 3rd and 4th, 1981 in Sioux Falls, South Dakota. Ames: Iowa State Univ. Press, 1983, 247pp.

Series of technical papers on effects of various kinds of radiation on cells and use of radiation for diagnosis and treatment. Contains some unusual papers on interactions of cells with light.

*U.N. Scientific Committee on Effects of Atomic Radiation (UNSCEAR). *Sources and Effects of Ionizing Radiation.* U.N. Publ. E-77-IX-1, 1977, 725pp.

Massive and authoritative U.N. report on all forms of radiation from

nuclear sources. Attempts to compile and interpret on international basis all
existing data on radiation sources and their effects.

*_____. *Ionizing Radiation Sources and Biological Effects.* U.N. Publ. E-82-
IX-8, 1982, 773pp.
    Most recent UNSCEAR report. Updates and expands data and in-
terpretations of the 1977 report.

U.S. Department of Energy and University of Texas School of Public Health.
Natural Radiation Environment III. Proceedings of a Symposium Sponsored by the U.S. Department of Energy and the University of Texas School
of Public Health, Houston, Texas, April 1978. 2 vols. Washington, D.C.:
U.S. Dep. Energy, Technical Information Center, 1980, 1736pp.
    Two volumes containing many papers on origins of natural radioactiv-
ity in world, particularly radon and other radioactive isotopes from uranium and thorium chains and cosmic radiation.

RADON

Cohen, Bernard L. "Radon: Characteristics, Natural Occurrence, Technological
Enhancement, and Health Effects." *Progress in Nuclear Energy,* 4, 1979.
    Good general brief review of radon problem.

National Council on Radiation Protection and Measurement. *Exposures from
the Uranium Series with Emphasis on Radon and Its Daughters.* NCRP
Report 77, 1984, 131pp.
    Discusses uranium distribution and its relationship to radon concentra-
tion. Recommendations on how to minimize exposure.

*_____. *Evaluation of Occupational and Environmental Exposures to Radon
and Radon Daughters in the United States.* NCRP Report 78, 1984, 204pp.
    Comprehensive study of sources of radon in United States and its im-
pact on humans. Contains extensive additional references.

O'Riordan, M. C., A. C. James, S. Rae, and A. D. Wrixon. *Human Exposure to
Radon Decay Products Inside Dwellings in the United Kingdom: A Memo-
randum of Evidence to the Royal Commission on Environmental Pollution.*
National Radiation Protection Board Report NRPB-R152, 1983, 41pp.
    Emphasizes extent of presence of radon inside dwellings in England
and discusses risks and methods for control.

U.S. Environmental Protection Agency. *Indoor Radiation Exposure Due to Ra-
dium-226 in Florida Phosphate Lands.* EPA 520/4–78–013. Feb. 1979,
198pp, and app.
    Discusses radon exposure levels in central Florida, the risks posed, and
possible ways to minimize them.

EFFECTS OF NUCLEAR WEAPONS

Adams, Ruth, and Susan Cullen, eds. *The Final Epidemic: Physicians and Scientists on Nuclear War.* Chicago: Educational Foundation for Nuclear
Science, 1981, 254pp.
    Series of articles on political, social, and medical aspects of nuclear
war.

*Committee for Compilation of Materials on Damage Caused by Atomic Bombs in Hiroshima and Nagasaki. *The Physical, Medical, and Social Effects of the Atomic Bombings.* Translated from the Japanese by Eisei Ishikawa and David L. Swain. New York: Basic Books, 1981, 706pp.

If you really want to know in human terms what a nuclear bomb does, read this book. Unemotional, factual account of what actually happened in two Japanese cities, and pictorial realities throughout text are more persuasive against nuclear weapons than any emotional texts yet written.

Ehrlich, Paul R., et al. "Long-Term Biological Consequences of Nuclear War." *Science* 222 (23 Dec. 1983):1293–1300.

Extension of the nuclear winter concept with particular emphasis on its ecological effects.

*Glasstone, Samuel, and Philip J. Dolan. *The Effects of Nuclear Weapons.* 3d ed. Washington, D.C.: Government Printing Office, 1977, 653pp.

Thorough and detailed explanation of nuclear explosions and their effects. Should be required reading for all educated citizens.

U.S. Congress. Office of Technology Assessment, Nuclear War Effects Project Staff. *The Effects of Nuclear War.* Washington, D.C.: Government Printing Office, 1979, 151pp.

Somewhat coldly analytical look at possible effects of nuclear war on United States or USSR. Includes possible effects of bomb on Detroit and a fictionalized account of what might happen if a bomb exploded over Charlottesville, Virginia.

U.S. Department of Energy. *Nuclear Proliferation and Civilian Nuclear Power,* Executive Summary, and *Proliferation Resistance.* Vol. 2. DOE/NE-001, 1980.

Huge 9 volume study aims to evaluate relationship between nuclear power and nuclear weapons to determine U.S. policy towards nuclear power in other nations.

Turco, R. P., O. B. Toon, T. P. Ackerman, J. B. Pollack, and Carl Sagan. "Nuclear Winter: Global Consequences of Multiple Nuclear Exposions." *Science* 222 (23 Dec. 1983). 1283–1292.

Provides primary technical basis for concept of "nuclear winter" as possible aftermath of multiple nuclear explosions.

NUCLEAR REACTORS FOR POWER GENERATION

*American Nuclear Society. *Report of the Special Committee on Source Terms.* La Grange Park, Ill.: American Nuclear Society, Sept. 1984. 457pp.

Detailed report by Special Committee on Source Terms, supported by Department of Energy, Nuclear Regulatory Commission, and Electric Power Research Institute. Discusses chemistry and escape paths of significant radioisotopes in a major nuclear accident.

*Glasstone, Samuel, and Walter H. Jordan. *Nuclear Power and Its Environmental Effects.* La Grange Park, Ill.: American Nuclear Society, 1980, 395 pp.

An excellent text for general public on nuclear reactors and their environmental aspects.

*U.S. Nuclear Regulatory Commission. *Reactor Safety Study.* WASH-1400, Oct. 1975, 198pp.

Main report, the so-called Rasmussen Report, estimates risks and consequences of nuclear reactor accidents (many appendices). Although controversial, it is still best and most objective analysis existing.

_____. *Reactor and Fuel Cycle Descriptions.* Vol. 9. DOE/NE-0001, 1980, 419pp. and app.

Thorough and comprehensive review of nuclear power reactor design and entire reactor fuel cycle. Covers uranium mining, chemical separation, isotope separation, fuel preparation, and waste disposal.

DISPOSAL OF NUCLEAR WASTE

*U.S. Department of Energy. *Management of Commercially Generated Radioactive Waste: Final Environmental Impact Statement.* Vol. 1. DOE/EIS00046F, Oct. 1980, 576pp.

Environmental impact statement drafted as first step in establishing national nuclear waste disposal policy. Considers all feasible methods of waste disposal and advantages and disadvantages of each. Provided much of technical basis for the Nuclear Waste Policy Act of 1982.

_____. *Proceedings of the 1981 National Waste Terminal Storage Program Information Meeting.* DOE/NWTS-15, Nov. 1981, 335pp.

Collection of papers on science and technology of underground waste disposal and associated political and social problems.

_____. *Long-Term Management of Liquid High-Level Radioactive Wastes Stored at the Western New York Nuclear Service Center, West Valley.* DOE/EIS-0081, June 1982, 154pp. and app.

Environmental impact statement aimed at much publicized cleanup of radioactive waste at abandoned reprocessing site at West Valley, N.Y. Contains all considerations required for any complex waste disposal situation.

_____. *Low-Level Radioactive Waste Policy Act Report: A Response to Public Law 96-573.* DOE/NE0015, 1981(?), 58pp.

Report issued by the Department of Energy as required by the Low Level Waste Policy Act of 1980. It covers future needs for disposal capacity, status of available sites, transportation requirements, and interim storage methods.

ELECTROMAGNETIC RADIATION: SOURCES AND EFFECTS

Barnes, P. R., E. F. Vance, and H. W. Askins, Jr. *Nuclear Electromagnetic Pulse (EMP) and Electric Power Systems.* Oak Ridge National Laboratory Report ORNL-6033, Apr. 1984, 61pp.

Excellent discussion of electromagnetic pulses resulting from a nuclear weapon explosion and their potential for causing major damage to our electrical power generation grid.

James F. MacLaren, Ltd. *A Report on the State-of-the-Art of the Effects of Transmission Lines: A Report to the Canadian Electrical Association.* Montreal: Canadian Electrical Association, 31 Mar. 1979, 283pp.

Excellent broad review, from utility viewpoint, of environmental impact of high-voltage transmission lines.

Johns Hopkins Applied Physics Laboratory. *Hazard Analysis for Magnetic In-*

duction from Electric Transmission Lines. Report PPSE T-23, June 1983, 64pp. and app.

Brief, well-done analysis of fields generated under high-voltage transmission lines and possible effects of fields on humans.

*National Council on Radiation Protection. *Radiofrequency Electromagnetic Fields: Properties, Quantities and Units, Biophysical Interaction, and Measurements.* NCRP Report 67, Mar. 1981, 134pp.

Highly scientific explanation of origins of electromagnetic radiation, its behavior, and laws governing its interaction with matter. Numerous equations and graphs.

National Institute of Environmental Health Sciences. *Proceedings of the US-USSR Workshop on Physical Factors: Microwaves and Low Frequency Fields.* Research Triangle Park, N.C.: National Institute of Environmental Health Sciences, Apr. 1981, 271pp.

Papers from joint U.S.-USSR workshop meeting held at Seattle, Washington, in 1979. An excellent comparison of views of two countries on effects of microwaves.

*Sheppard, Asher R., and Merrill Eisenbud. *Biological Effects of Electric and Magnetic Fields of Extremely Low Frequency.* New York: New York Univ. Press, 1977, 255pp.

Critical review of technical literature on effects of electrical and magnetic fields from 0 to 300 Hertz (covering frequencies at which electric power is generated, transmitted, and used). Factual, thorough, and complete. Leaves interpretation to reader.

"Symposium on Health Aspects of Nonionizing Radiation." *Bulletin New York Academy Medicine* 55, No. 11 (Dec. 1979), 334pp.

Collection of papers from symposium covering biological effects of microwaves and public issues concerning them.

*World Health Organization. *Nonionizing Radiation Protection.* WHO Regional Publ. European Ser. 10, 1982, 267pp.

Thorough review from international viewpoint of sources and effects of all kinds of nonionizing radiation. Contains extensive lists of references. For technically educated reader.

Steneck, N. H., H. J. Cook, A. J. Vander, and G. L. Kane. "The Origins of U.S. Safety Standards for Microwave Radiation." Science 208 (13 June 1980): 1230–1237.

Interesting analysis of origins of U.S. microwave standards and motivations and behavior of people involved in setting them.

POPULAR TEXTS

A large number of books aimed at the general public have been written about radiation, nuclear power, and nuclear weapons. Unfortunately many of them have been written to support a particular position, pro- or antinuclear for example, or have tried to dramatize hazards. The dramatic sells well. The following list is a selection of the best written and most interesting, regardless of position. Let reason and fairness guide your judgement.

Brodeur, Paul. *The Zapping of America: Microwaves, Their Deadly Risk and the Cover-Up.* New York: W. W. Norton, 1977.

Book that first called public attention to the possible hazards of microwaves. Dramatic, well written, and intentionally inflammatory, it suggests deliberate suppression by military-industrial complex of possible public risk from microwaves.

Cohen, Bernard L. *Before It's Too Late: A Scientist's Case FOR Nuclear Energy.* New York: Plenum Press, 1983.

Probably the best-written and most effective statement of pronuclear position yet.

Harvard Nuclear Study Group. *Living with Nuclear Weapons.* Cambridge: Harvard Univ. Press, 1983.

Excellent, realistic look at the societal and political problems created by nuclear weapons, and their possible solutions.

Hurley, Patrick M. *Living with Nuclear Radiation.* Ann Arbor: Univ. Michigan Press, 1982.

Brief, balanced, well-written and factual book that avoids taking sides.

McCracken, Samuel. *The War Against the Atom.* New York: Basic Books, 1982.

Vigorous defense of nuclear power. Strongly attacks antinuclear movement and many of the people leading it.

Panati, Charles, and Michael Hudson. *The Silent Intruder: Surviving the Radiation Age.* Boston: Houghton Mifflin, 1981.

Question and answer format. Emphasizes radiation hazards and is strongly opposed to nuclear power.

Shapiro, Fred C. *Radwaste: A Reporter's Investigation of a Growing Nuclear Menace.* New York: Random House, 1981.

Thorough, well-researched and well-written discussion of the problems of nuclear waste disposal. Emphasizes the problems more than the solutions.

Steneck, Nicholas H. *The Microwave Debate.* Cambridge, Mass: MIT Press, 1985.

Carefully researched and thorough review of origins and nature of controversies surrounding issue of standards for microwave exposure. Contains cogent thoughts on relationship between science and public. Don't read Brodeur unless you read this too.

Sternglass, Ernest. *Secret Fallout: Low Level Radiation from Hiroshima to Three Mile Island.* New York: McGraw-Hill, 1981.

Personalized book that tries to make the case that low levels of radiation are much more harmful than establishment is willing to admit. Strongly antinuclear.

INDEX

223

226